Extension
Physics

Bryan Milner

CAMBRIDGE
UNIVERSITY PRESS

Series Editor   Bryan Milner
Author   Bryan Milner
Consultant   Sam Ellis

PUBLISHED BY THE PRESS SYNDICATE OF THE UNIVERSITY OF CAMBRIDGE
The Pitt Building, Trumpington Street, Cambridge, United Kingdom

CAMBRIDGE UNIVERSITY PRESS
The Edinburgh Building, Cambridge CB2 2RU, UK
40 West 20th Street, New York, NY 10011–4211, USA
10 Stamford Road, Oakleigh, Melbourne 3166, Australia
Ruiz de Alarcón 13, 28014 Madrid, Spain
Dock House, The Waterfront, Cape Town 8001, South Africa

http://www.cambridge.org

First published 1998
Reprinted 2000, 2001

Printed in the United Kingdom at the University Press, Cambridge

Typeface Stone Informal 10/13.5 pt

A catalogue record for this book is available from the British Library

ISBN 0 521 64917 X paperback

Designed and produced by Gecko Limited, Bicester, Oxon

Cover photograph: Pole vault; M. Brett/Telegraph Colour Library

■ Acknowledgements

11t, Ecoscene; **11b**, Ecoscene/Anthony Cooper; **30**, photo courtesy
of HR Wallingford Ltd; **66**, **69cl**, **69cr**, Andrew Lambert; **35**, Chris
Priest/Science Photo Library; **41**, Rev. Ronald Royer/Science Photo
Library; **46**, Crown copyright.

Contents

How to use this book

■ An introduction for students and their teachers

The basic *Science Foundations* text-books for Biology,
Chemistry and Physics contain all that you need to know
for GCSE Science tests and examinations at the
Foundation level.

This *Science Foundations Extension* book contains all of the
extra physics that you need to know for GCSE Science
tests and examinations at the Higher level.

There are some differences in the way that you should use
this *Science Foundations Extension* book compared with the
way that you used the basic *Science Foundations* books.
If you are to gain maximum benefit from this book, it
is important that you know what some of these
differences are.

A box at the beginning of each double-
page *Extension* spread will tell you which
Science Foundations topics you should
have completed before you start. It may
also remind you of some of the scientific
ideas from those *Science Foundations*
topics.

At the end of each section of a double-
page spread you will find a question, or
sometimes two questions. These questions
have been carefully designed to cover all
the things you need to know for Higher
tier GCSE tests and examinations.

Make sure you answer the questions as
clearly and as fully as you can. They will
then build up into a very useful set of
revision notes.
Your answers to these questions will, in
effect, be your summaries of the
important ideas on each spread.
This means that there is no need for any
'What you need to remember' passages
for you to copy and complete as you did
in *Science Foundations*.

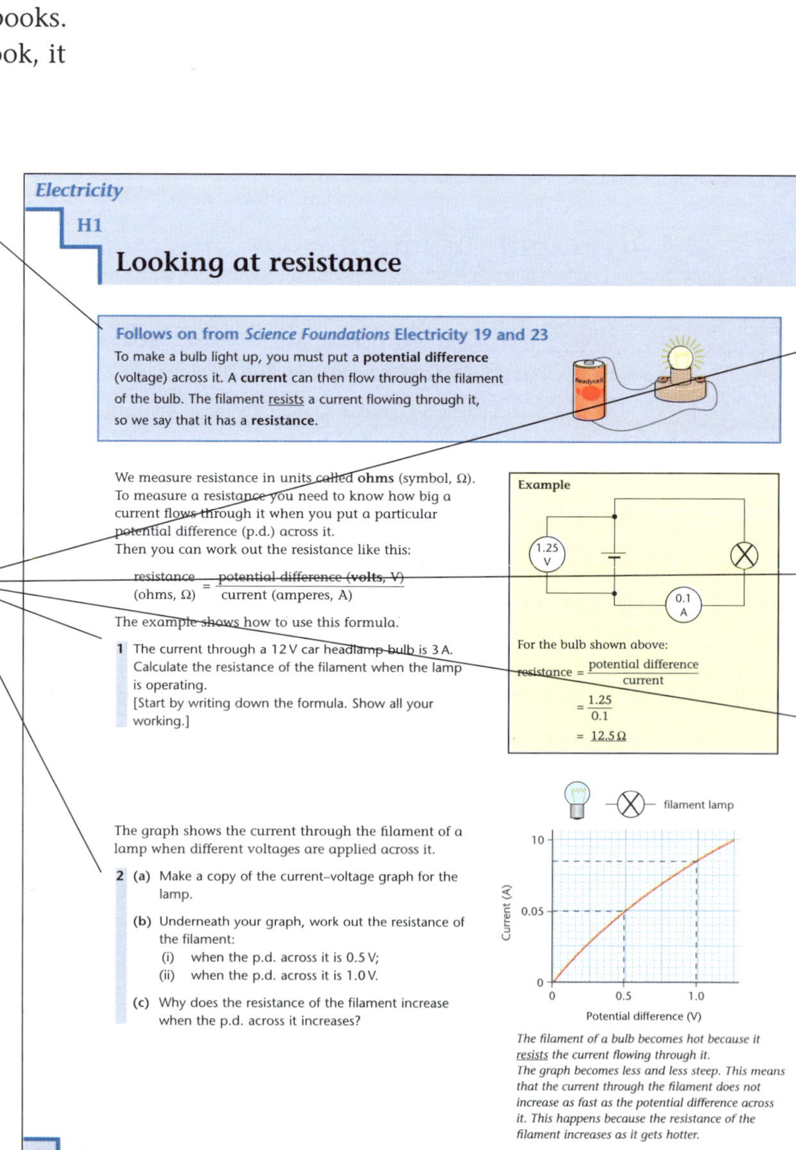

This graph shows the current through a **resistor** that stays at a constant temperature.

3 (a) Sketch the graph. [You don't need graph paper; it's just the shape you need to get right.]

(b) Copy and complete the sentence.

For a resistor at constant temperature, the current increases in proportion to the applied ___________.

The graph is a straight line through the origin (0, 0). This means that the current is directly proportional to the voltage. This happens because the resistance stays the same.

The resistance of some electrical components depends on their surroundings. A **thermistor**, for example, is designed so that its resistance changes a lot when the temperature changes.

4 (a) Sketch the resistance–temperature graph for the thermistor.

(b) Describe how the resistance of the thermistor changes with temperature.

A **light dependent resistor (LDR)** is designed so that its resistance depends on the intensity of the light that falls on it.

5 (a) Sketch the resistance–light intensity graph for the light dependent resistor (LDR).

(b) Describe how the resistance of the LDR changes with the light intensity.

Using your knowledge

EI 1 Work out the resistance of the resistor shown at the top of the page. [You can do this using any voltage.]

EI 2 What is the resistance:

(a) of the thermistor at 75°C;

(b) of the LDR in the dark?

EI 3 A potential difference of 230 V is applied across a resistance of 920 Ω. Calculate the current.

EI 4 What does this current–voltage graph tell you about the resistance of a diode?

You will find some further questions in the 'Using your knowledge' section at the end of each double-page spread. To answer these questions, you need to <u>use</u> the ideas from the spread to solve problems, to explain new situations or to help you to interpret data. Many of the questions in Higher tier GCSE tests and examinations expect you to be able to use your scientific knowledge in this way. The correct answers to these questions are included at the back of this *Extension* book so that you can check out your own answers. You need to find the reference number at the beginning of a question and then look up this reference number in the Answers to 'Using your knowledge' questions.

Finally, you will also find, at the back of this book:
- some 'Additional questions' for each module to give you further practice in answering the sorts of questions that you will be asked in tests and examinations;
- some advice about how you can make sure that you gain as many marks as you can in Higher tier examination papers;
- a list of the physics formulas that you need to know and/or to be able to use;
- a 'Glossary/index' that tells you the meanings of the most important scientific words that you meet in this *Extension* book and the pages on which you can find these words being used.

Explaining conduction, convection and radiation

> **Follows on from *Science Foundations* Energy 1–4**
>
> Energy is transferred from hotter places to cooler places in three ways:
>
> - by **conduction** – energy is transferred through a substance, usually a solid, but the substance itself does not move;
> - by **convection** – energy is transferred when a hot liquid or gas moves;
> - by **radiation** – the energy is carried by electromagnetic waves, which do not need any substance to travel through.

Explaining conduction

Metals, when they are solid or liquid, are good **conductors** of **thermal energy**. Most other substances are poor thermal conductors.

Gases are poor conductors of thermal energy. Even metals, when they are hot enough to be gases, are poor thermal conductors. This means that <u>all</u> gases are good thermal <u>insulators</u>.

The diagrams explain why solid and liquid metals are good conductors and why gases are always poor conductors.

1. **(a)** Which substances are good thermal conductors?

   (b) What is special about the structure of metals that makes them good conductors?

   (c) Explain how the free electrons in metals enable them to transfer thermal energy by conduction.

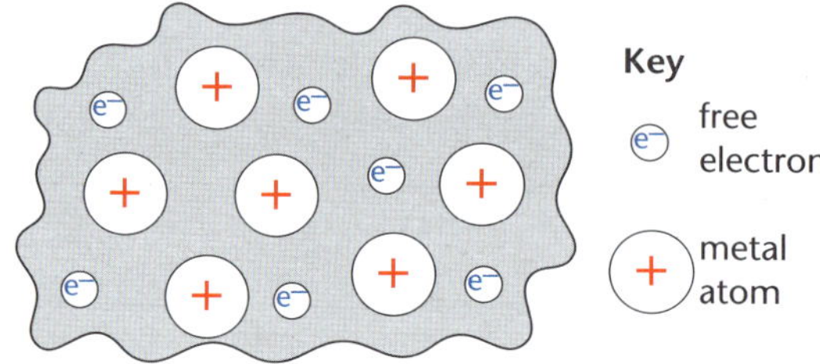

*The atoms in metals have some **electrons** that can move about anywhere in a piece of metal. Heating the metal gives these electrons more kinetic energy. They carry this energy with them as they travel through the metal. They also transfer energy to other electrons when they collide with them.*

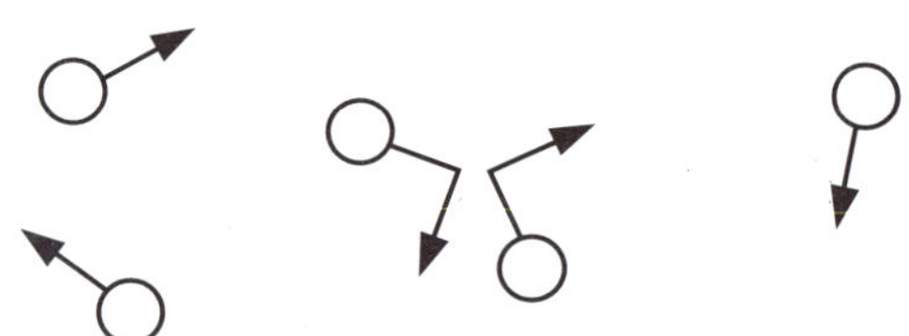

The particles in a gas move about and collide with each other. But they don't move as fast as electrons. Also there are a lot fewer particles in each cm^3 than there are in a solid or a liquid. So gases are poor thermal conductors.

Explaining convection

Convection occurs because liquids and gases become less dense when they are warmer. To understand why this happens, you need to imagine a 'bubble' of liquid completely surrounded by the same liquid.

The 'bubble' of liquid doesn't move because the forces that act on it are <u>balanced</u>. The weight of the liquid pulling downwards is balanced by the force of the surrounding liquid pushing upwards.

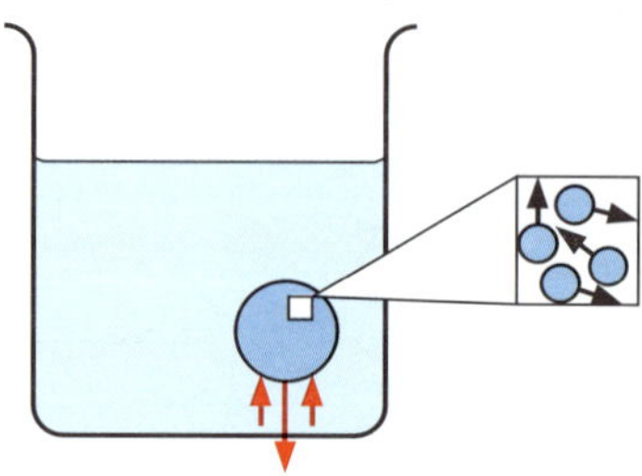

If the 'bubble' of liquid is heated, energy is transferred to the particles. The particles move about faster; they have more **kinetic energy**. This means that they take up more space. So the 'bubble' of liquid expands. It still has the same mass so the weight of the 'bubble' is still the same. But the 'bubble' has a bigger volume, so the surrounding liquid pushes up on it more.

There is now an unbalanced upwards force, so the 'bubble' moves up. Colder liquid moves in to take its place. This movement is called a **convection current**.

2 Copy and complete the sentences.
As you heat up a 'bubble' of a liquid or a gas, the particles gain more ___________ energy.
The particles take up more ___________ so the liquid or gas becomes less ___________.
The upwards force of the surrounding liquid or gas on the bubble ___________, but there is no change in the ___________ of the bubble. So the bubble of warm liquid or gas moves ___________. This movement is called a ___________ current.

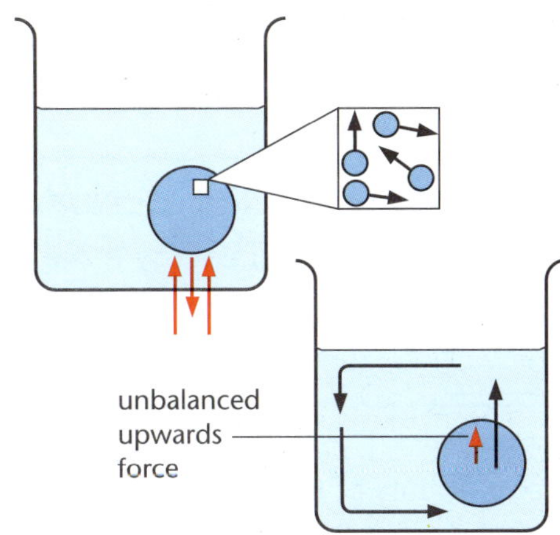

The same <u>mass</u> of particles is now in a larger volume. So the 'bubble' of liquid is now less <u>dense</u> than it was before.
Exactly the same thing happens when a gas is heated.

Explaining radiation

All objects which are above the absolute zero of temperature (–273 °C) give out (emit) electromagnetic radiation. The <u>rate</u> at which an object radiates energy and the <u>wavelength</u> of the radiation it emits both depend on how hot the object is.

The hot mug emits long wavelength infrared radiation. You can <u>feel</u> it with your skin, but you can't <u>see</u> it.

The <u>very</u> hot filament of the lamp emits shorter wavelength radiation that you can see.

Using your knowledge

En 1 The graphs show how the radiation emitted by an object depends on its temperature. Describe, as fully as you can, what the graphs tell you.

(a)   (b)

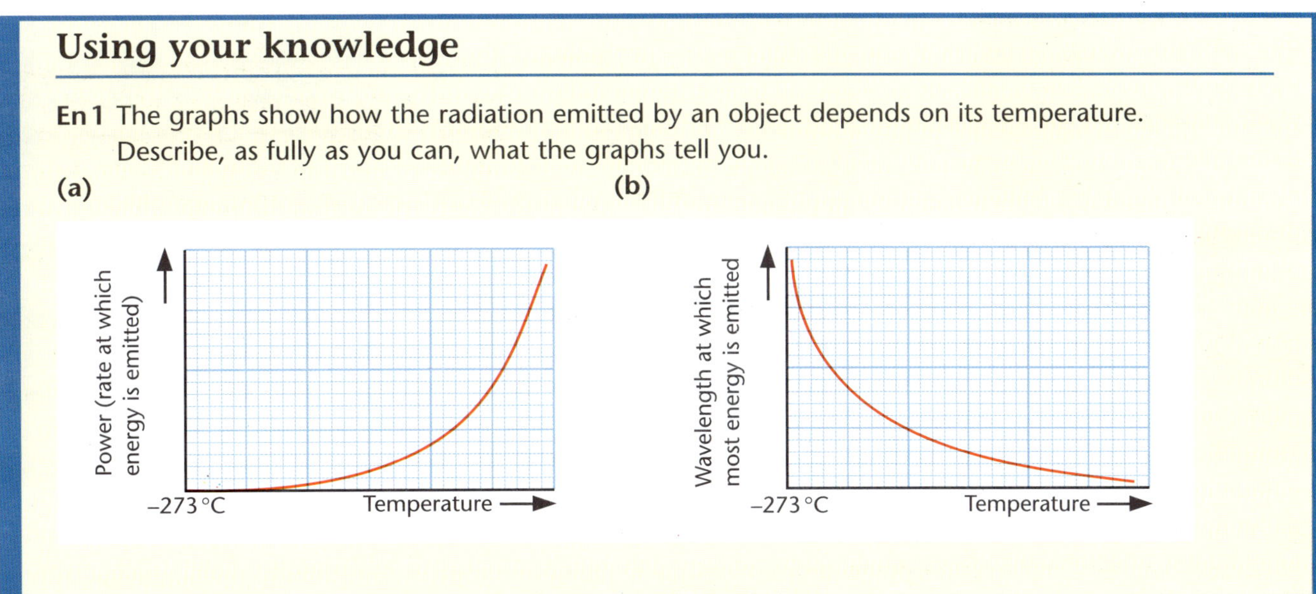

How soon will it pay for itself?

Heat is lost, i.e. **thermal energy** is transferred, from warm buildings to the colder surroundings outside. We can reduce the rate of this energy transfer in many ways, including double glazing, loft insulation, cavity wall insulation and draughtproofing.

We can do lots of things to buildings to reduce thermal energy transfer. But before we spend money on things like double glazing or loft insulation, we need to know how much money this will save on heating bills.

When we are deciding if it is worth spending money on a particular improvement, it is useful to know how long it will take for the improvement to pay for itself. This is called the **pay-back time**. The shorter the pay-back time, the more **cost-effective** the improvement is.

1 (a) Some ways of reducing thermal energy transfer are more cost-effective than others.
Explain, as fully as you can, what this means.

(b) Write down the energy-saving ideas on this page in order of their cost-effectiveness.

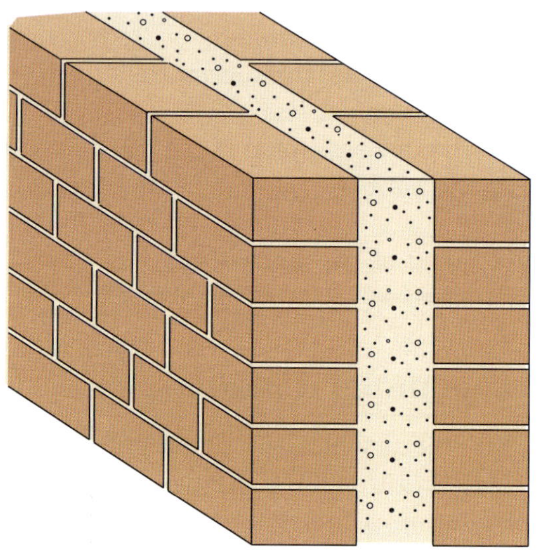

Cavity wall insulation   Costs £450
Saves £75 per year

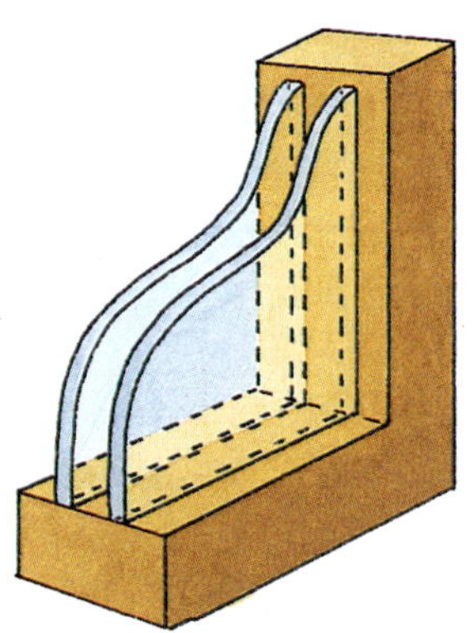

Double glazing   Costs £1200
Saves £60 per year

Draught excluders   Costs £25
Saves £50 per year

Loft insulation   Costs £150
Saves £150 per year

Using your knowledge

En 2 Fifty years ago, most houses in the UK were heated using coal fires.

Nowadays, most houses are heated using gas fires or gas central heating.

(a) Look at the diagrams. Then copy and complete the following.

Cost of installing gas fire = _____________

Annual saving in fuel = _____________

Pay-back time = _____________

(b) How cost-effective is installing the gas fire compared with methods of reducing thermal energy transfer from a house?

(c) What other advantages do gas fires have over coal fires?

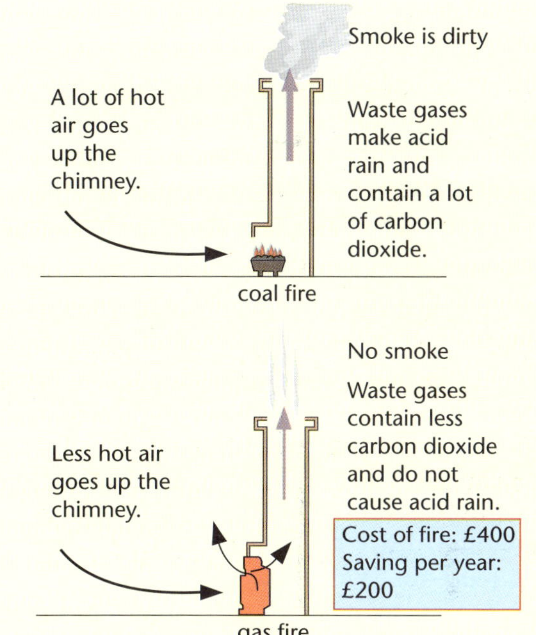

It is easy to control the thermal energy output of a gas fire. It is also very easy to turn it on or off.

En 3 Some houses are 'all-electric'. This means that electricity is used for all the heating and cooking as well as for lighting, TV sets, etc.

So no fuel is burned inside the house. But to keep the air inside the house fresh, it needs to be changed about once an hour. In older houses this happens through cracks, for example around doors or between floorboards. Newer houses can be built with hardly any cracks and with a ventilation system fitted.

(a) Explain how a heat exchanger reduces the thermal energy loss from a house.

(b) What would the pay-back time be for the extra cost of reducing cracks and fitting a heat exchanger?

(c) How cost-effective would you say that the heat exchanger is?

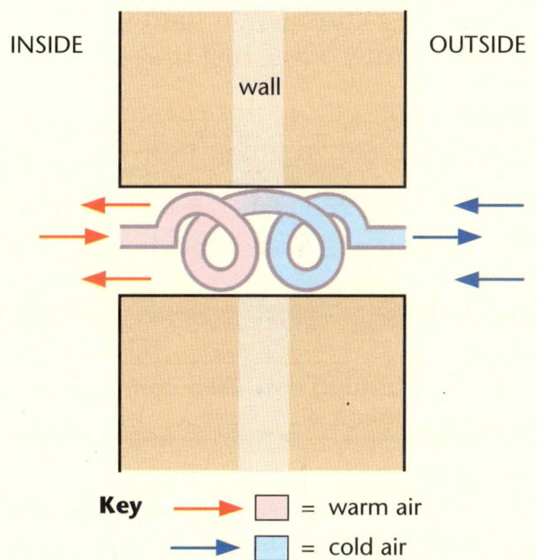

A 'crack-free' house with a heat exchanger costs about £2000 more to build.

It can save about £400 a year in fuel bills.

How should we generate our electricity?

<u>Cost</u> is another important factor when deciding what energy sources to use to generate electricity.

There are several different costs to take into account. For example, the energy in the wind is free, so wind generators have no fuel costs. But the energy in the wind is spread out quite thinly. So to generate the same amount of electricity as a coal-fired power station you need a very large number of wind generators. It costs more to build all these wind generators than it does to build a coal-fired power station.

To work out what it costs to produce each Unit (kilowatt-hour, kWh) of electricity, we have to work out:

- the cost of <u>building</u> the system (the capital cost);
- the cost of <u>operating</u> and <u>maintaining</u> the system;
- the cost, if any, of the <u>fuel</u> that is used.

We then need to share this cost between all the Units of electricity the system produces in its lifetime.

1 **(a)** Copy and complete the table for electricity generated by wind generators, by a nuclear power station, by a gas-fired power station and by a coal-fired power station.

System	Building cost (per kWh)	Fuel cost (per kWh)	Operating cost (per kWh)	Total cost (per kWh)

(b) Which system has

   (i) the lowest building cost,
   (ii) the lowest fuel cost,
   (iii) the lowest overall cost

for each kWh of electricity?

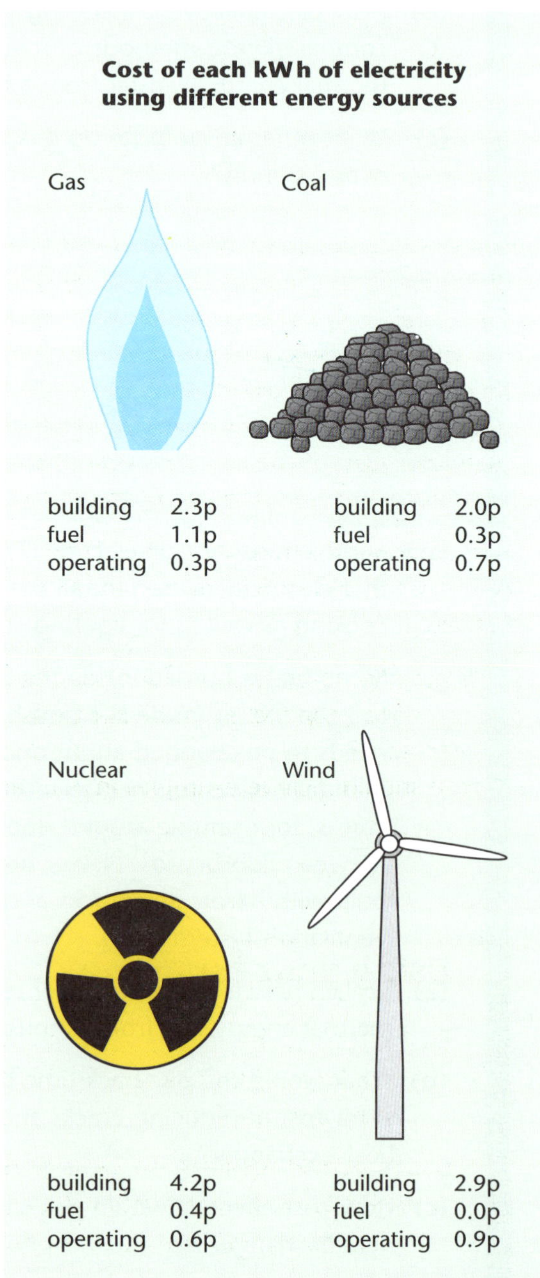

Electricity generating systems eventually wear out.
They then have to be dismantled and removed.
This **de-commissioning** costs money and should also be
taken into account when calculating the cost of each
Unit of electricity.

2 Does de-commissioning increase the cost of each Unit
of electricity more for wind generators or for nuclear
power stations?

*Nuclear power stations don't produce waste gases
that pollute the air, but they do produce solid waste
that will stay dangerously radioactive for thousands
of years. We don't really know how much it will cost
to store this safely for such long periods of time, or
even whether it will be possible to do so. The power
stations themselves contain a lot of radioactive
material so dismantling them is expensive.*

Generating electricity also has other 'costs' that it is
difficult to put a figure on. For example, burning some
fossil fuels helps to create acid rain and the damage this
causes costs money. Burning all types of fossil fuel
produces carbon dioxide which may increase the
greenhouse effect. This may change the climate and
raise the level of the sea. The cost of this could be
enormous but is very difficult to estimate.

Some people think that we should tax things that
damage the environment to help to pay for the damage
caused.

3 What other factors should we take into account when
comparing the 'costs' of different ways of generating
electricity?
You should include examples in your answer.

*Many people care about the appearance of their
surroundings. They would be prepared to pay more
for their electricity to avoid, for example, having
lots of wind generators (wind-farms) on most
hill-tops.*

Using your knowledge

En 4 Which of the energy sources shown
on page 10 do <u>not</u> contribute to the
greenhouse effect.

En 5 A person speaking on TV says that wind
generators produce no pollution.
Do you agree? Explain your answer.

En 6 When nuclear power stations are
operating properly, they produce very
little pollution.
Why do you think that many people are
opposed to them?

Matching electricity supply and demand

During each 24-hour period, the demand for electricity rises and falls quite a lot. Sometimes there is a sudden big increase in demand – for example, when many people switch on their electric kettles at the same time. This can happen at the end of a very popular TV programme or at half-time during a televised cup final.

The electricity generating companies have to generate the right amount of electricity to meet these sudden increases in demand. But they don't want to generate this amount of electricity <u>all</u> the time because it isn't all needed so it won't all be sold. To solve this problem, they need to have some generators that they can stop and start very quickly.

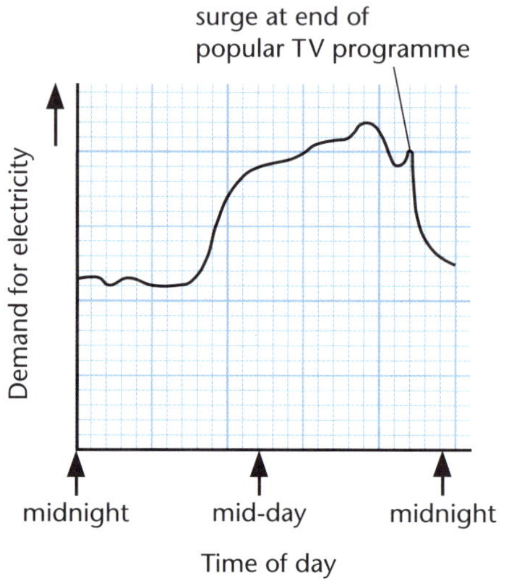

1 (a) Which type of power station can be stopped and started most quickly to meet sudden changes in demand?

(b) Which type of power station is least suited to meet sudden changes in demand?

Because nuclear power stations take a very long time to start up and shut down, they are used to produce much of the electricity that is <u>always</u> needed. This is called the **base-load demand**. Other types of power station are started up when they are needed. But there are still times, especially at night, when too much electricity is generated. Some of this surplus electricity is sold very cheaply to any customers who want it, but some of it is used to prepare for sudden increases in demand.

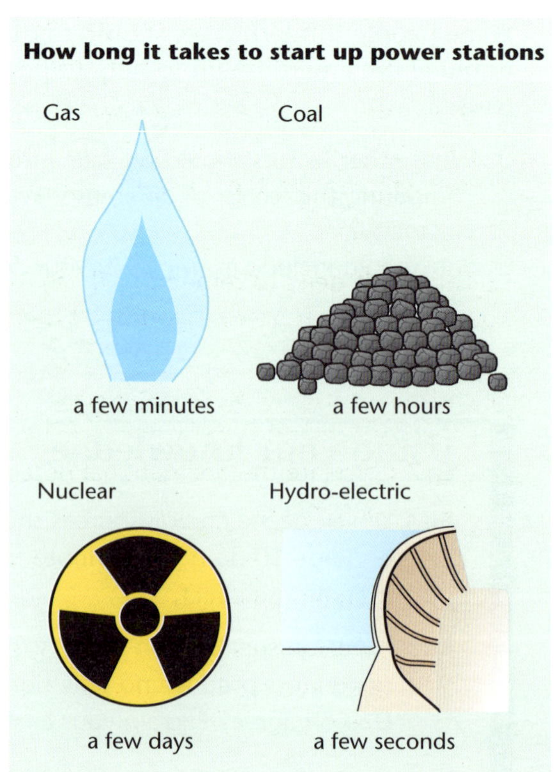

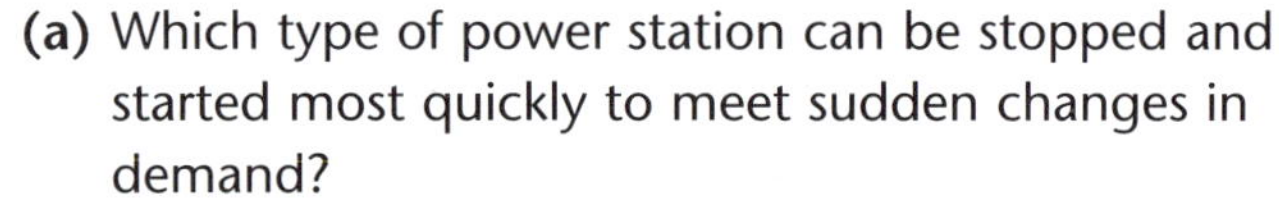

Surplus electricity can't be stored, but it can be used to pump water uphill to a storage reservoir. The water from this reservoir can then be used to generate electricity when there is a sudden increase in demand. This system is called a **pumped storage system**.

2 Explain, as fully as you can, the idea behind a pumped storage system.

Pumped storage system

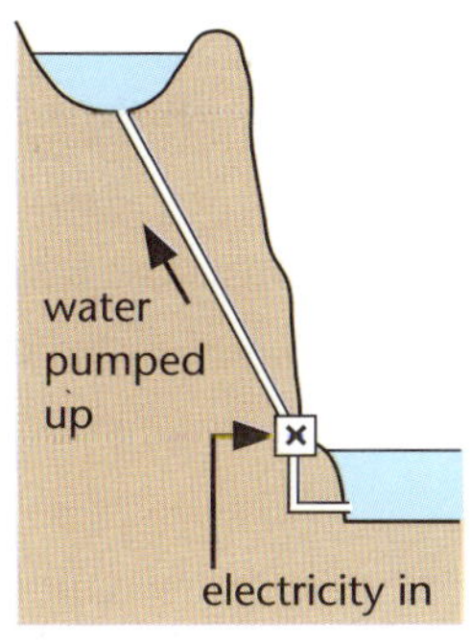

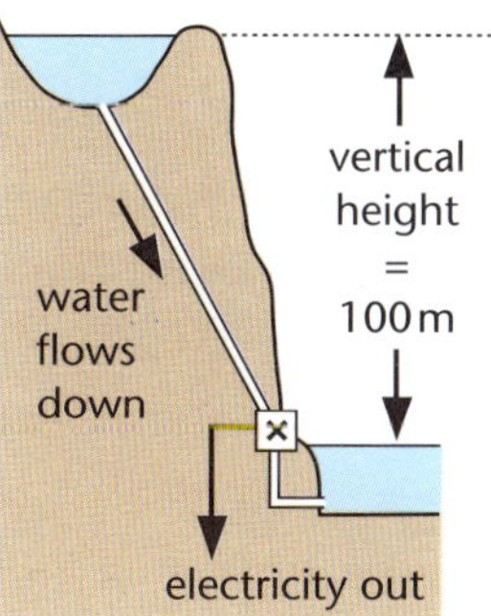

■ Calculating gravitational potential energy

You can calculate the energy stored in something that is lifted up like this:

* $$\text{gravitational potential energy (J, joules)} = \text{weight (N, newtons) } [10 \times \text{mass in kg}] \times \text{vertical height raised (m, metres)}$$

3 Calculate the energy transferred to a 25 kg sack of potatoes when it is lifted up 1.5 m on to the back of a lorry.

Example

The pumped storage system shown above lifts 1 tonne of water 100 metres. So the gravitational potential energy transferred to this water is:

$$(10 \times 1000)\,\text{N} \times 100\,\text{m} = 10\,000\,\text{N} \times 100\,\text{m}$$
$$= \underline{1\text{ million J}}$$

■ Calculating efficiency

You can calculate the efficiency of an energy transfer like this:

* $$\text{efficiency} = \frac{\text{useful energy transferred by device}}{\text{total energy transferred to device}}$$

4 To lift the 25 kg sack of potatoes on to the lorry, a person's muscles transfer 17 750 J of energy. How efficient are the muscles?

[* You don't need to remember these formulas. You will be given them if you need them.]

Example

The pump uses 1.25 million J of energy to lift the 1 tonne of water 100 m.

So the efficiency of the transfer is:

$$\frac{1\text{ million J}}{1.25\text{ million J}} = \underline{0.8}\ (\times 100 = \underline{80\%})$$

Using your knowledge

En 7 Calculate the gravitational potential energy:

(a) transferred to a 1 kg mass when you lift it from the floor on to a shelf 2 m high;

(b) transferred to the body of a 50 kg girl when she climbs 1500 m to the top of a mountain.

En 8 Calculate the efficiency:

(a) of the pumped storage generator if it generates 750 000 J of electricity when 1 tonne of water falls 100 m;

(b) of the pumped storage system overall.

Looking at resistance

Follows on from *Science Foundations* Electricity 19 and 23

To make a bulb light up, you must put a **potential difference** (voltage) across it. A **current** can then flow through the filament of the bulb. The filament <u>resists</u> a current flowing through it, so we say that it has a **resistance**.

We measure resistance in units called **ohms** (symbol, Ω). To measure a resistance you need to know how big a current flows through it when you put a particular potential difference (p.d.) across it.
Then you can work out the resistance like this:

$$\text{resistance (ohms, Ω)} = \frac{\text{potential difference (\textbf{volts}, V)}}{\text{current (amperes, A)}}$$

The example shows how to use this formula.

1 The current through a 12 V car headlamp bulb is 3 A. Calculate the resistance of the filament when the lamp is operating.
[Start by writing down the formula. Show all your working.]

Example

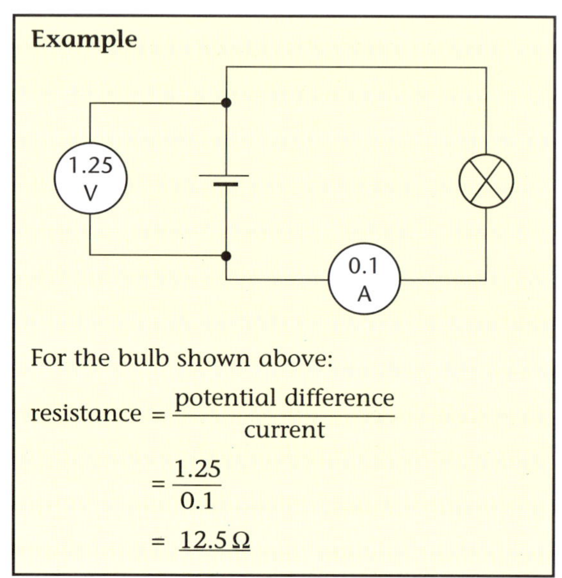

For the bulb shown above:

$$\text{resistance} = \frac{\text{potential difference}}{\text{current}}$$

$$= \frac{1.25}{0.1}$$

$$= \underline{12.5\ Ω}$$

The graph shows the current through the filament of a lamp when different voltages are applied across it.

2 (a) Make a copy of the current–voltage graph for the lamp.

(b) Underneath your graph, work out the resistance of the filament:
   (i)   when the p.d. across it is 0.5 V;
   (ii) when the p.d. across it is 1.0 V.

(c) Why does the resistance of the filament increase when the p.d. across it increases?

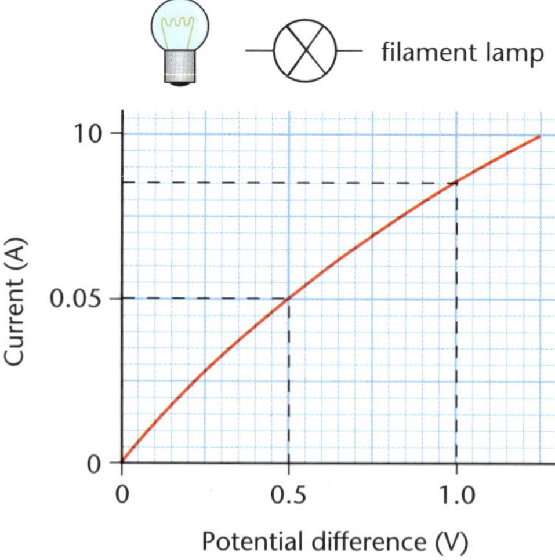

The filament of a bulb becomes hot because it <u>resists</u> the current flowing through it.
The graph becomes less and less steep. This means that the current through the filament does not increase as fast as the potential difference across it. This happens because the resistance of the filament increases as it gets hotter.

This graph shows the current through a **resistor** that stays at a constant temperature.

3 (a) Sketch the graph. [You don't need graph paper; it's just the shape you need to get right.]

(b) Copy and complete the sentence.

For a resistor at constant temperature, the current increases in proportion to the applied ___________.

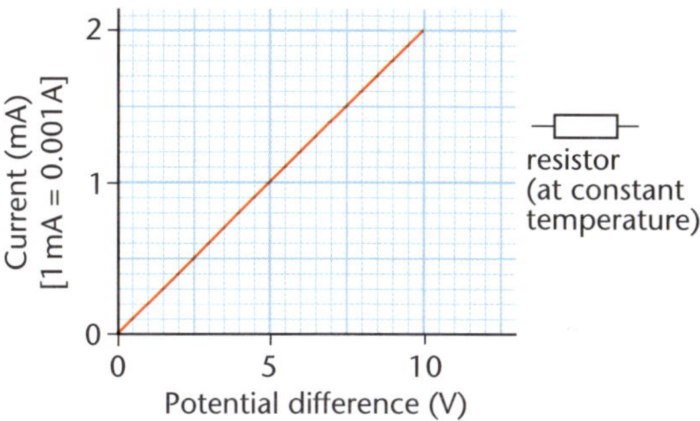

The graph is a straight line through the origin (0, 0). This means that the current is directly proportional to the voltage. This happens because the resistance stays the same.

The resistance of some electrical components depends on their surroundings. A **thermistor**, for example, is designed so that its resistance changes a lot when the temperature changes.

4 (a) Sketch the resistance–temperature graph for the thermistor.

(b) Describe how the resistance of the thermistor changes with temperature.

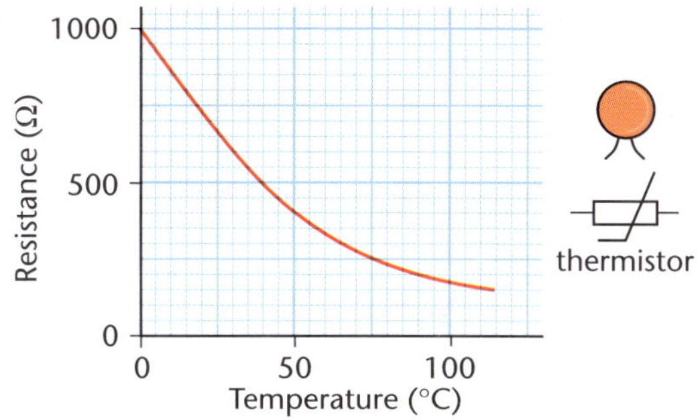

A **light dependent resistor** (**LDR**) is designed so that its resistance depends on the intensity of the light that falls on it.

5 (a) Sketch the resistance–light intensity graph for the light dependent resistor (LDR).

(b) Describe how the resistance of the LDR changes with the light intensity.

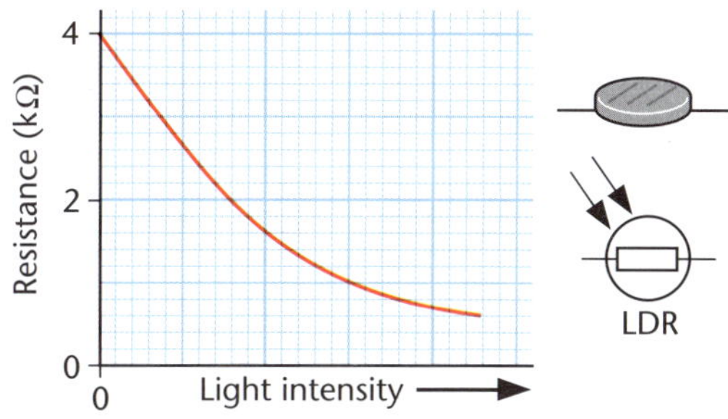

Using your knowledge

EI 1 Work out the resistance of the resistor shown at the top of the page. [You can do this using any voltage.]

EI 2 What is the resistance:

(a) of the thermistor at 75°C;

(b) of the LDR in the dark?

EI 3 A potential difference of 230 V is applied across a resistance of 920 Ω. Calculate the current.

EI 4 What does this current–voltage graph tell you about the resistance of a diode?

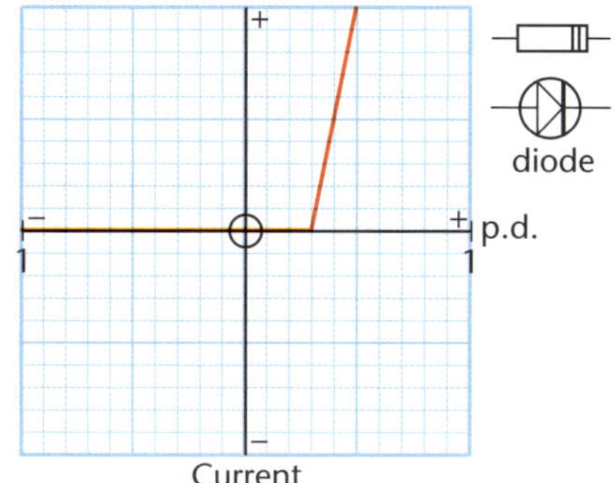

More about generators

In a small generator used on a bicycle, a magnet spins between coils of wire.

In the generator shown in the diagram, a coil of wire rotates in a magnetic field.

1 List <u>four</u> factors that affect the size of the potential difference that is produced by a generator of this kind.

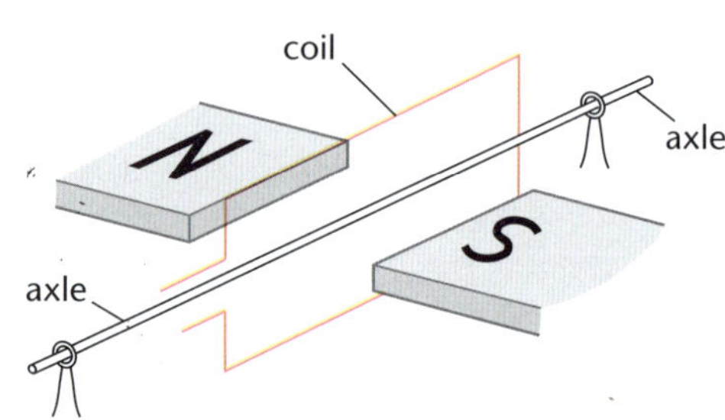

The size of the potential difference induced across the ends of the coil can be increased by:
- *using stronger magnets;*
- *using more turns on the coil;*
- *spinning the coil faster;*
- *using a coil with a bigger area.*

Rotating the coil rather than the magnet causes a problem if you want to <u>use</u> the induced potential difference. You can't just connect a circuit across the ends of the coil because the wires will become twisted. The diagram shows how this problem is solved.

2 Describe <u>in words</u> how the potential difference induced across the ends of a generator coil can be applied to a circuit.

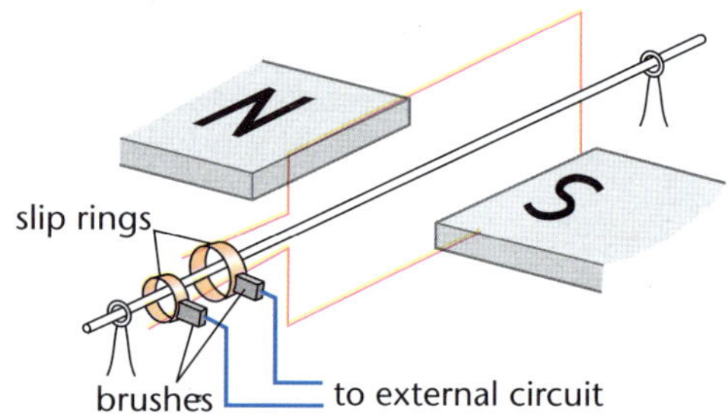

*The **brushes** are made of graphite which conducts electricity but is soft so it doesn't wear away the **slip rings**.*

Large commercial generators, such as those used in power stations, usually have a rotating <u>electro</u>magnet called a **rotor**. The alternating current is induced in fixed coils called **stators**.

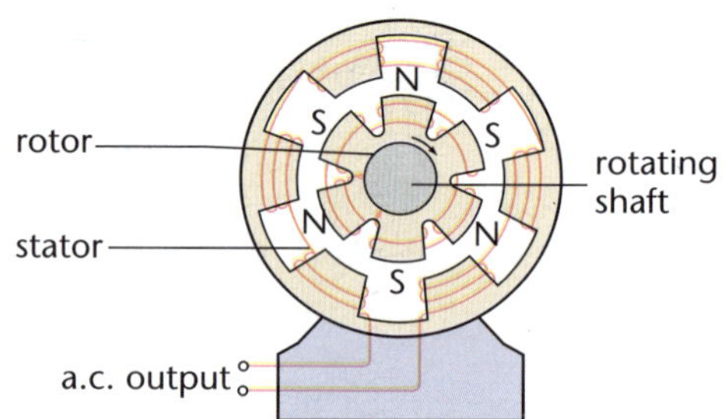

Using your knowledge

E15 (a) The electromagnets used in commercial generators are very powerful.
Why is this an advantage?

(b) The electricity for the electromagnets is often produced by the generator itself.
Suggest a problem with this idea.

How transformers work

The diagram shows the main parts of a transformer. It also explains how an alternating current input to the transformer produces an output.

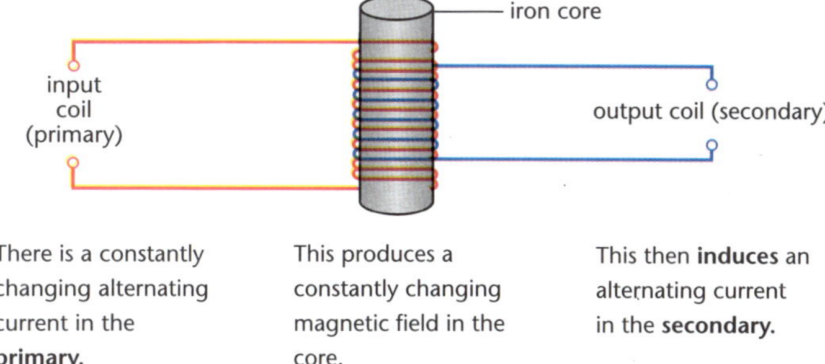

There is a constantly changing alternating current in the **primary.**

This produces a constantly changing magnetic field in the core.

This then **induces** an alternating current in the **secondary.**

1 Explain how there can be an alternating current in the secondary coil of a transformer even though it is not directly connected to a power supply.

A transformer can step the voltage up or down because it has a different number of turns on its primary and secondary coils. In fact:

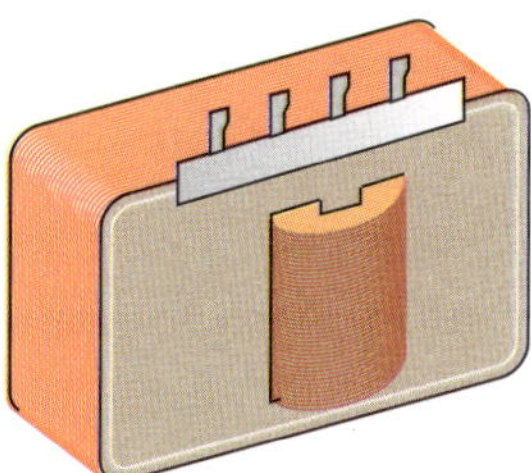

A real transformer looks like this.
The core is made of layers (it is laminated).
It also goes around the outside of the coils as well as through the middle of them.

$$\frac{\text{secondary voltage}}{\text{primary voltage}} = \frac{\text{number of secondary turns}}{\text{number of primary turns}}$$

The example in the box uses this formula.

2 A 230 V mains transformer gives an output of 4.6 V for a radio. There are 2500 turns on the primary coil of the transformer. How many turns are there on the secondary coil?
[Write down the formula and show your reasoning.]

> **Example**
> A 230 V mains transformer used in a TV set has 100 turns on its primary coil. It produces a p.d. of 4600 V across its secondary.
>
> The p.d. is stepped up $\frac{4600}{230} = \underline{20 \text{ times}}$.
>
> So there are 20 times more turns on the secondary, i.e. $20 \times 100 = \underline{2000 \text{ turns}}$.

Using your knowledge

E16 A transformer at a power station steps up the voltage from 25 000 V to 400 000 V.

(a) What does this tell you about the primary and secondary coils of the transformer?

(b) What would be the output voltage of the transformer if you used it to step up a 230 V input voltage?

H4

Transmitting mains electricity efficiently

Follows on from *Science Foundations* Electricity 6, 22 and *Extension* Electricity H3

Power is the rate at which energy is transferred:

power	=	**potential difference**	×	**current**
(watts, W)		(volts, V)		(amperes, A)

Mains electricity is stepped up to a very high voltage before leaving the power station. It is then stepped down again to a safer voltage before being used. This is done so that less energy is wasted heating up the transmission line cables.

You can get a higher <u>voltage</u> out of a transformer than you put into it. But you can't get more <u>energy</u> out of it than you put in.

The rate at which energy is transferred to and from the transformer is given by:

power	=	p.d.	×	current
(watts, W)		(volts, V)		(amperes, A)

The power can't become bigger, so if the voltage <u>increases</u> 100 times for example, the current must <u>decrease</u> 100 times.

1 At a power station the voltage is increased from 25 000 to 400 000 V.
What happens to the current?

The smaller the current the less the power that is wasted in the cables. The reduction in the wasted power is <u>more than</u> in proportion to the reduction in the current (see the example in the box).

2 How many times smaller are the power losses in transmission cables when the voltage is increased from 25 000 to 400 000 V?

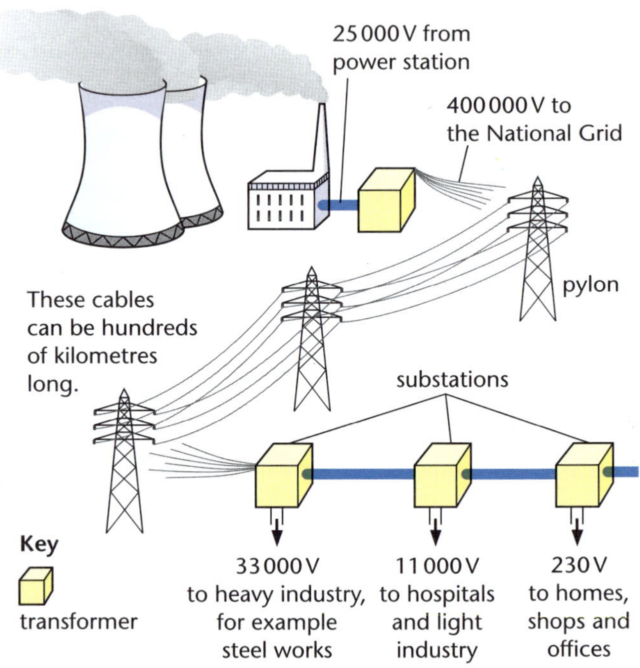

Example
If you step <u>up</u> the voltage 10 times, you step <u>down</u> the current 10 times.
This reduces the **power loss** $10 \times 10 = \underline{100\ times}$.
In other words, by stepping up the voltage 10 times, you reduce the power loss to 1% of what it was before.

Using your knowledge

EI7 The mains supply in your home is 230 V. Suppose it was transmitted at this voltage.

(a) How much bigger would the current be in the transmission cables?

(b) How many times greater would the power losses be?

H5 Why mains appliances with metal cases need earthing

Follows on from *Science Foundations* Electricity 4 and *Extension* Electricity H4

An electrical circuit normally contains a **fuse**.

If a current becomes too large, the fuse melts and breaks the circuit.

Mains electricity is supplied through the **live** (brown) and **neutral** (blue) wires. There is hardly any potential difference between the neutral wire and the **earth** wire (green and yellow).

The graph shows the potential difference between the live wire and the neutral wire during a small fraction of a second.

1 Describe how the potential difference between the live wire and the neutral wire varies with time.

The potential difference between the live wire and the earth wire varies in exactly the same way. So if a live wire touches an unearthed metal case and you then touch the metal case, a dangerous current will flow through your body to the Earth.

To make appliances with metal cases safe, we earth them. The diagram shows what then happens if the live wire touches the earthed metal case of an appliance.

2 Explain, in as much detail as you can, why earthing the metal case of a mains appliance makes it safer.

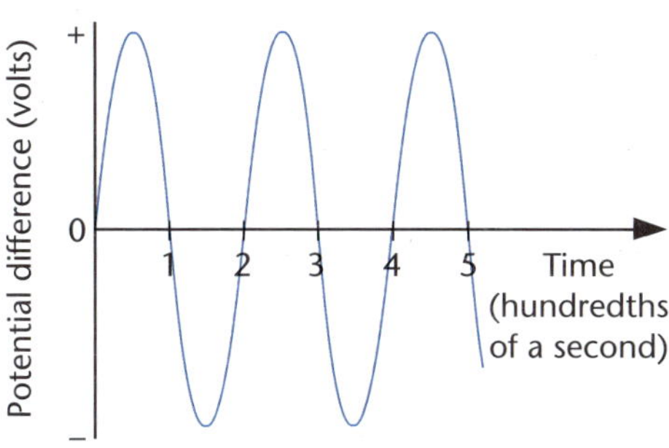

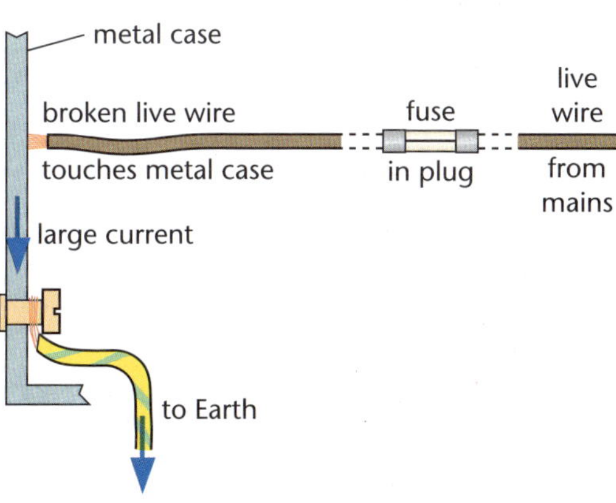

If a broken live wire touches an earthed metal case, a large current flows to Earth. This current flows through the fuse. More thermal energy is transferred by the fuse so it becomes hotter and melts. This disconnects the live wire from the metal case.

Using your knowledge

EI8 A fuse has only a small resistance so it transfers only a very small amount of electricity as thermal energy.

(a) What happens to the rate of thermal energy transfer if the current through the fuse increases 1000 times?

(b) How does this explain what happens to the fuse?

> **REMEMBER**
> The rate of energy transfer in a resistance is in proportion to the current <u>squared</u>.

A closer look at electrical charge

> **Follows on from *Science Foundations* Electricity 22 and 24**
>
> An electric **current** is a flow of electric **charges**.
>
> In metals (and in graphite), it is negatively charged **electrons** that move.
>
> In molten or dissolved ionic compounds, the charged particles that move are called **ions**.
>
> Some ions have a **positive** charge; other ions have a **negative** charge.

The diagrams below explain how a metal such as
copper can conduct electricity.

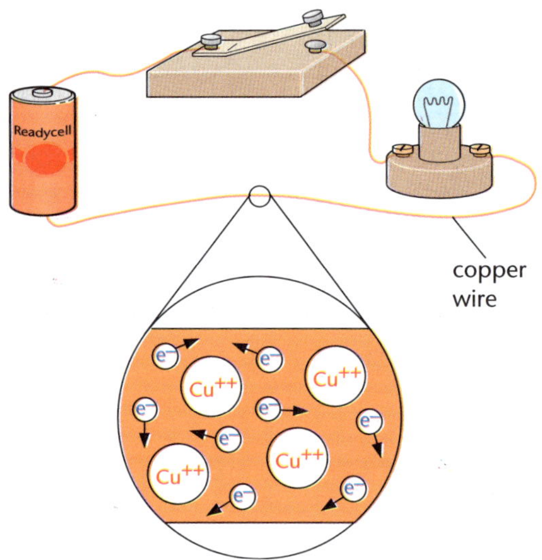

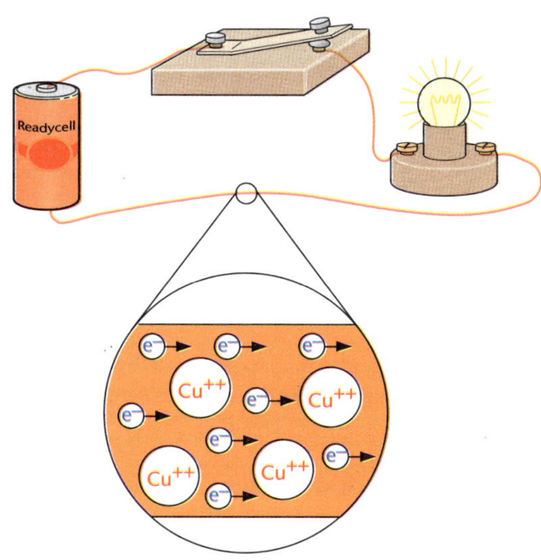

Key

 Cu^{++} copper atom that has lost electrons e^- 'free' electron that can move anywhere in the metal

A piece of copper does not have an overall electrical
charge. The positive and negative charges <u>balance</u>.
It is just that the free electrons don't belong to any
particular atom.

1 Explain <u>in words</u> why a metal like copper is a good
conductor of electricity.

The diagram on the right shows how a solution of
copper chloride conducts electricity.

2 Explain, as fully as you can <u>in words</u>, what is happening
when an electric current flows through copper chloride
solution.

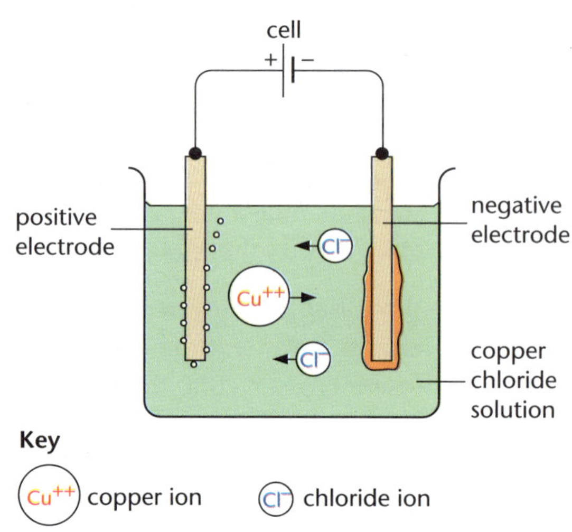

Key

Cu^{++} copper ion Cl^- chloride ion

When an electric current passes through copper chloride
solution, copper and chlorine are produced.
This process is called **electrolysis**.

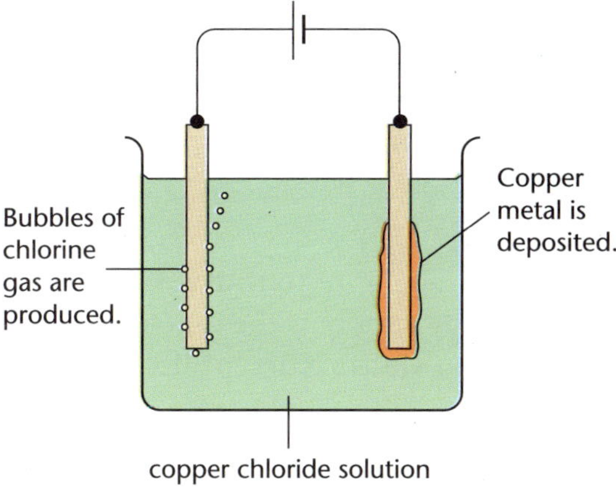

3 (a) <u>Where</u> are copper and chlorine produced during
the electrolysis of copper chloride?

(b) What <u>two</u> factors does the <u>amount</u> of copper or
chlorine produced during electrolysis depend on?

*The amount (mass or volume) of copper and
chlorine produced is proportional to:*

- *the size of the current;*
- *the time that the current flows.*

The amount of charge that flows while an electric
current is switched on also depends on the size of the
current and how long it is on. We measure charge in
units called **coulombs** (C, for short).

The amount of charge that flows is given by:

$$\begin{array}{ccccc}
\text{charge} & = & \text{current} & \times & \text{time} \\
\text{(coulombs, C)} & & \text{(amperes, A)} & & \text{(seconds, s)}
\end{array}$$

So the mass (or volume) of copper and chlorine
produced during the electrolysis of copper chloride is
also directly proportional to the number of coulombs of
charge that has flowed.

> **Example**
> If you use a current that is half as big but
> for ten times as long, you will produce
> $\frac{1}{2} \times 10 = 5$ times as much copper and
> chlorine.

> **Example**
> During an electrolysis, a current of
> 1.5 amperes is used for 5 minutes.
>
> $$\begin{aligned}
> \text{charge (C)} &= \text{current (A)} \times \text{time (s)} \\
> &= 1.5 \times (5 \times 60) \\
> &= \underline{450\,C}
> \end{aligned}$$
>
> [Remember to change
> the time to seconds.]

4 Work out the charge that flows when a current of 0.8 A
is switched on for 12 minutes.
[Start by writing down the formula. Show all your
working.]

Using your knowledge

EI 9 During an electrolysis, 20 cm^3 of
hydrogen gas is produced. How
much hydrogen will be produced if
a current twice as big is used for five
times as long?

EI 10 During an electrolysis, a current of
0.5 A flows for 2 hours and deposits
8.0 g of silver.

(a) How much charge has flowed?

(b) How much silver will be deposited if a
current of 0.2 A flows for $1\frac{1}{2}$ hours?

A closer look at voltage

Follows on from *Extension* Electricity H6

The voltage or **potential difference** (p.d.) of an electricity supply is the amount of 'push' it can give to make an electric **current** flow around a circuit.

1 Copy and complete the sentences.

   Potential difference is measured in units called __________.
   We measure potential difference using a ____________.

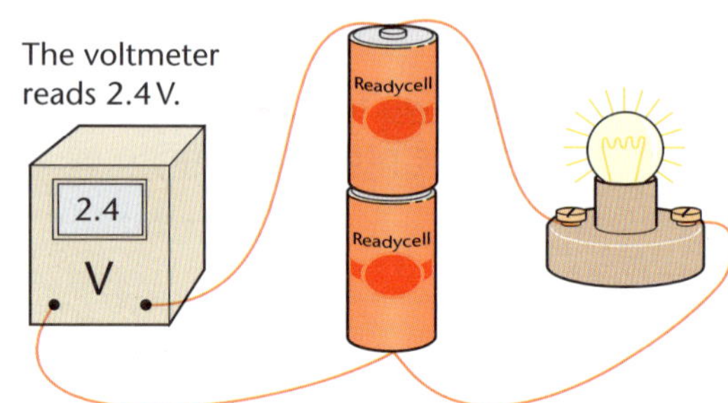

◼ Potential differences in static electricity

Batteries and power supplies can supply an electric current because they have a potential difference across their terminals.

You also get a potential difference between a charged object and the Earth. The graph shows how this potential difference varies with the size of the charge on the object.

2 Describe the relationship between the size of the electric charge on an object and the potential difference between the object and the Earth.
   [Make use of <u>all</u> the information from the graph.]

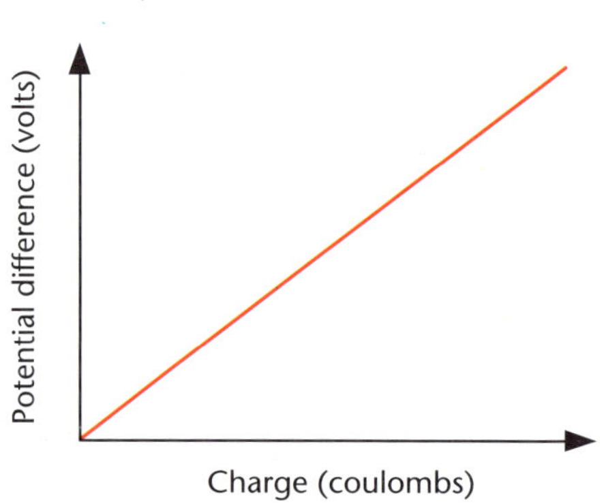

The diagrams show what can happen if the potential difference between an object and the Earth becomes big enough.

3 Explain why an electrically charged body can create a spark.
   [Make use of <u>all</u> the information from the diagrams.]

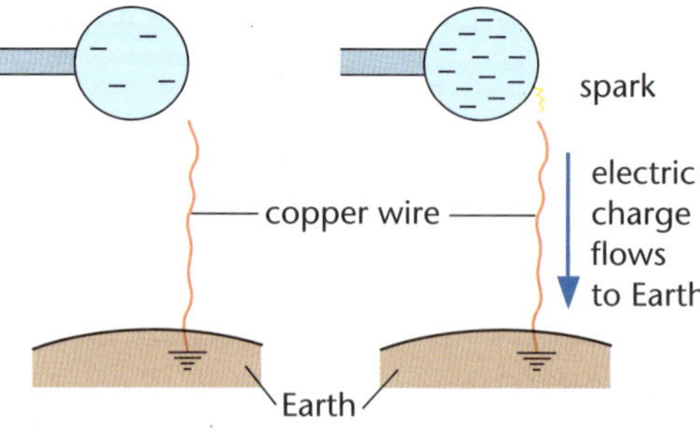

The same thing happens on a large scale with lightning.

What exactly is a volt?

An electric current is a flow of **charge**. When the charge flows through a resistance, energy is transferred.

The voltage or potential difference across a resistance is a measure of the energy that is transferred when a charge flows through it. If the flow of 1 coulomb of charge transfers 1 joule of energy, then the potential difference is 1 **volt**.

In other words:

energy transferred = potential difference ×   charge
   (joules, J)   (volts, V)   (coulombs, C)

[You don't have to remember this formula.
You will be given it if you need it.]

But [see the Remember box]: charge = current × time

so

    energy transferred = p.d. × current × time

This is why

   rate of energy transfer   =   p.d.   ×   current
      (J/s)   (volts, V)   (amperes, A)

[The rate of energy transfer
is the same as **power**
(watts, W).]

REMEMBER

When a current of one ampere is switched on for one second, one coulomb of charge flows:

    charge (C) = current (A) × time (s)

Example

The potential difference between a charged object and the Earth is 25 000 V. When the object is earthed, one millionth of a coulomb of charge flows.

   Energy transferred = p.d. × charge
                = 25 000 × 0.000 001
                = 0.025 J

Example

When 1000 C of charge flows, 400 kJ of energy are transferred. What is the potential difference that made the charge flow?

   Energy transferred = p.d. × charge

$$\text{So p.d.} = \frac{\text{energy transferred}}{\text{charge}}$$

$$= \frac{400\,000}{1000}$$

$$= 400\,\text{V}$$

Using your knowledge

EI 11 Calculate the following. [In each case start with the formula you are using. Show all your working.]

(a) Calculate the energy transferred when a potential difference of 12 V is used to make 500 C of charge flow.

(b) Calculate the potential difference used if the flow of 240 C of charge transfers 9600 J of energy.

EI 12 Explain why the power of an electrical appliance (in watts) is given by the formula:

    p.d.   ×   current
  (volts)   (amperes)

H1

Two types of wave

The diagram shows waves moving along a rope. Each wave is a disturbance that travels along the rope, but the rope itself does <u>not</u> travel along.

1 (a) In which direction are the waves travelling along the rope in the diagram?

(b) Describe how each <u>bit</u> of the rope moves as the waves pass along it.

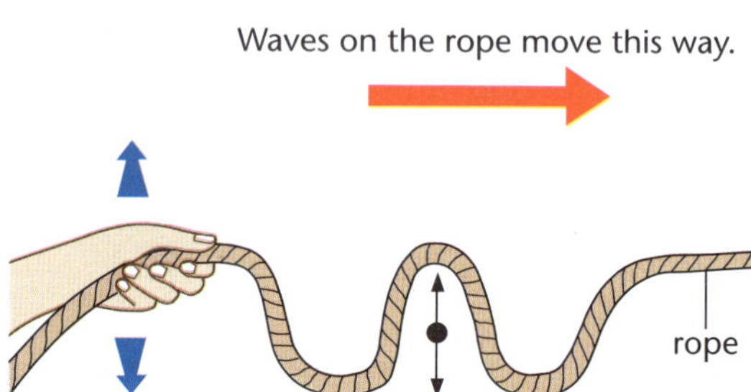

Because each bit of the rope vibrates <u>at right angles</u> to the direction that the waves are travelling, we say that the waves are **transverse** waves. ('Transverse' means 'across'.)

The diagram shows waves moving along a spring. Each wave is a disturbance that travels along the spring, but the spring itself does <u>not</u> travel along.

2 (a) In which direction are the waves travelling along the spring in the diagram?

(b) Describe how each <u>bit</u> of the spring moves as the waves pass through it.

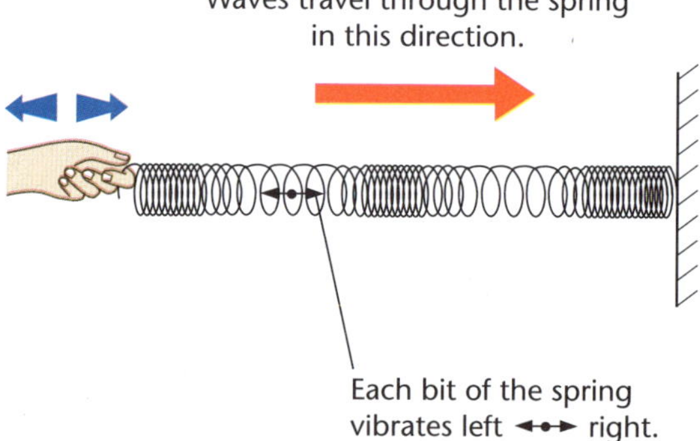

As the waves travel along the spring, each bit of the spring vibrates. These vibrations are <u>along the same direction</u> as the waves are travelling. So we say that the waves are **longitudinal** waves.

The diagrams show water waves and sound waves.

3 What types of wave are:

(a) water waves;

(b) sound waves?

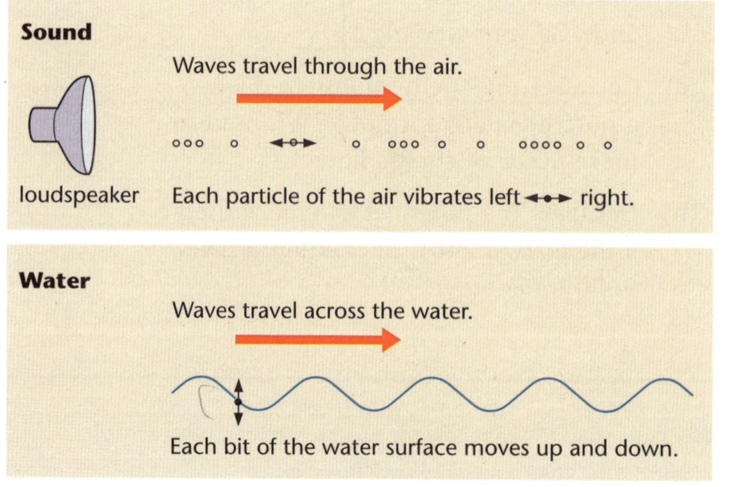

■ A formula for waves

The wave-maker shown in the diagrams has a **frequency** of 10 **hertz** (Hz). This means that it makes 10 waves every second. If you look at the diagrams, you will see that by the end of one second the first wave has travelled a distance equal to 10 wavelengths. So the speed of the waves is 10 wavelengths per second.

If the wave-maker is changed to a frequency of 4 Hz, the wave-speed is now 4 wavelengths per second.

In fact, for _any_ frequency:

wave-speed	=	frequency	×	wavelength
(metres per second, m/s)		(hertz, Hz)		(metres, m)

4 **(a)** A sound with a frequency of 440 Hz has a wavelength in air of 0.75 m.
Calculate the speed of sound in air.
[Start by writing down the formula. Show all your working.]

(b) In water, the 440 Hz sound travels at 1500 m/s. What is the wavelength of the sound in water?
[Show all your working.]

■ What type of wave are light waves?

Light waves are **electromagnetic waves**.

All electromagnetic waves are transverse waves.
They do not need a substance (**medium**) to travel through.
They can travel through empty space (a _vacuum_).

5 Write down _one_ difference between light waves and water waves.

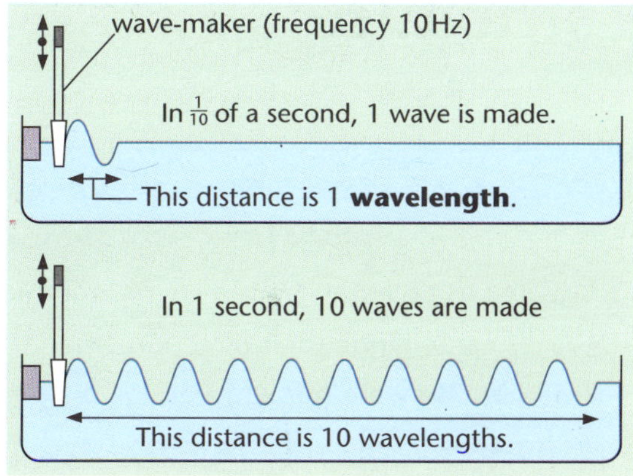

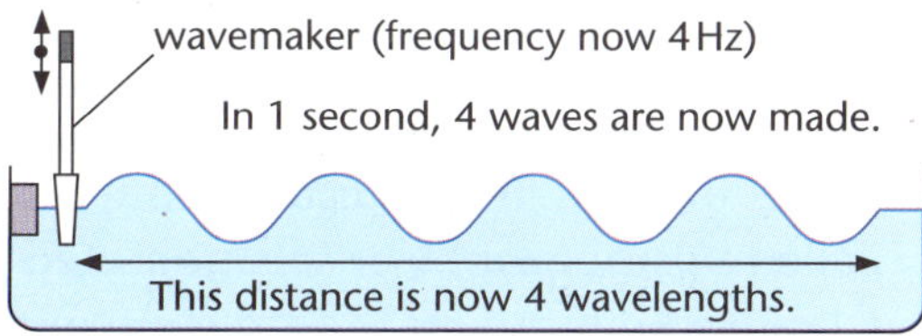

Example 1
Some water waves have a wavelength of 5 cm and a frequency of 4 Hz.
What is the wave-speed?

$$\text{wave-speed} = \text{frequency} \times \text{wavelength}$$
$$= 4 \times 5$$
$$= \underline{20\,\text{cm/s}} \quad (= 0.2\,\text{m/s})$$

Example 2
A high-pitched sound made by a dolphin has a frequency of 5000 Hz. The sound travels through the water at 1500 m/s.
What is the sound's wavelength?

$$\text{wave-speed} = \text{frequency} \times \text{wavelength}$$

So:

$$\text{wavelength} = \frac{\text{wave-speed}}{\text{frequency}}$$
$$= \frac{1500}{5000}$$
$$= \underline{0.3\,\text{metres}} \quad (= 30\,\text{cm})$$

Using your knowledge

W1 All radio waves travel at the same speed through the air. The wavelength of Radio 4 is 1500 m on long wave but on FM it is only 3 m.

(a) How do the frequencies of the long wave and FM transmissions compare?

(b) The frequency of the FM waves is 100 million hertz (100 MHz).
Calculate the speed of the radio waves.

W2 Blue light has double the frequency of red light. What does this tell you about the wavelengths of blue and red light?

H2 Using waves to find out about the Earth's structure

Earthquakes produce shock waves called
seismic waves. These can be detected at other
places on the Earth's surface using instruments
called **seismographs** or **seismometers**. Most of
what we know about the structure of the Earth
comes from studying seismic waves.

There are two types of seismic wave:
    primary waves or **P-waves** and
    secondary waves or **S-waves**.
The diagrams provide information about
these waves and what they tell us about the
structure of the Earth.

1 (a) What type of wave are:
    (i) P-waves;
    (ii) S-waves?

(b) Which travel faster, P-waves or S-waves?

(c) Which of the two types of seismic wave
    cannot travel through liquid? Give a
    reason for your answer.

2 Copy and complete the table.

Part of Earth	Thickness (km)	Solid or liquid?
crust		
mantle		
outer core		
inner core		

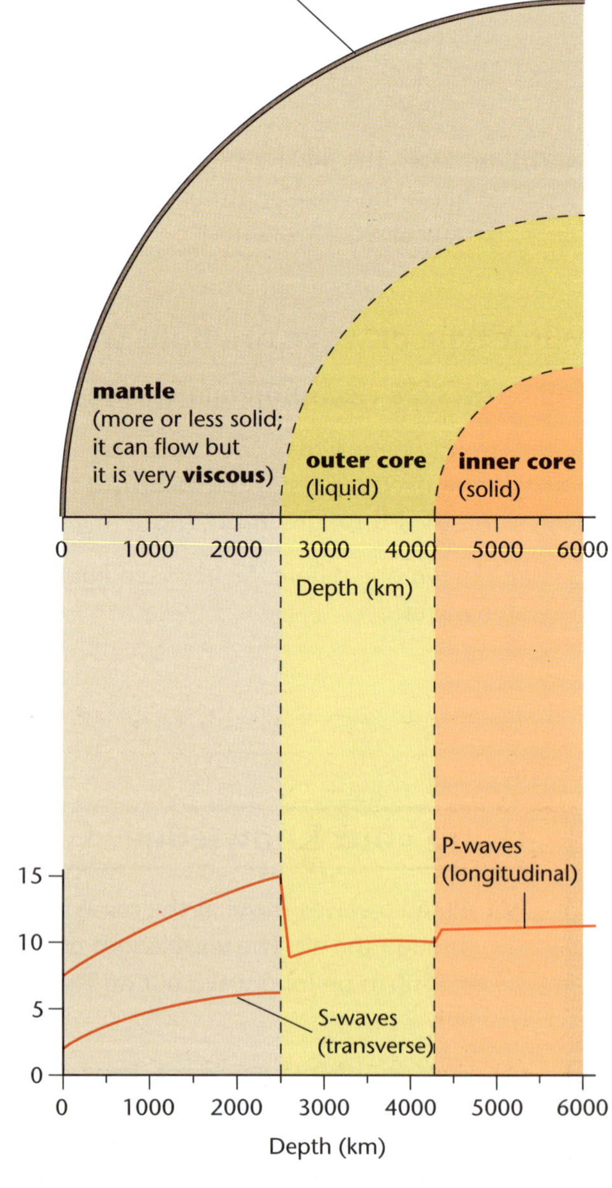

How S-waves travel through the Earth

As the mantle gradually becomes denser, the speed of S-waves gradually increases. This means that they are gradually refracted, as shown on the diagram.

3 (a) Make a copy of the diagram.

(b) Add a caption to your diagram explaining the curved path of the S-waves.

4 Why do S-waves produce a seismograph shadow zone?

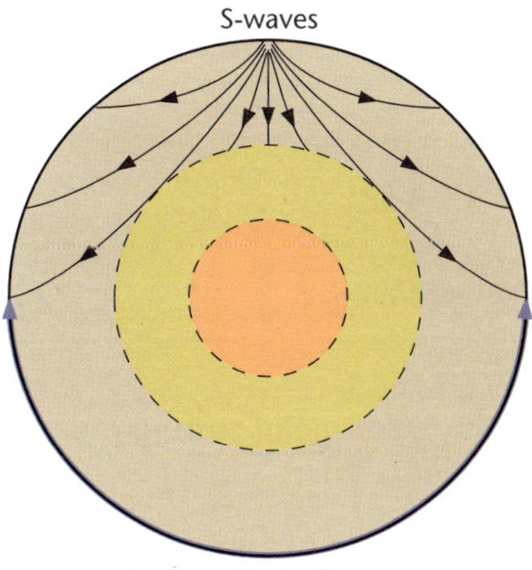

seismograph <u>shadow zone</u>

How P-waves travel through the Earth

P-waves are refracted as they travel through the Earth's mantle in just the same way as S-waves, but the P-waves also travel through the Earth's core.

Whenever there are sudden changes in the speed of the P-waves, refraction produces a sudden change in their direction.

4 (a) Make a copy of the diagram.

(b) Where do the sudden changes in the speed and direction of the P-waves mainly occur?

(c) Which P-wave changes its speed as it passes through the Earth but does not change its direction?

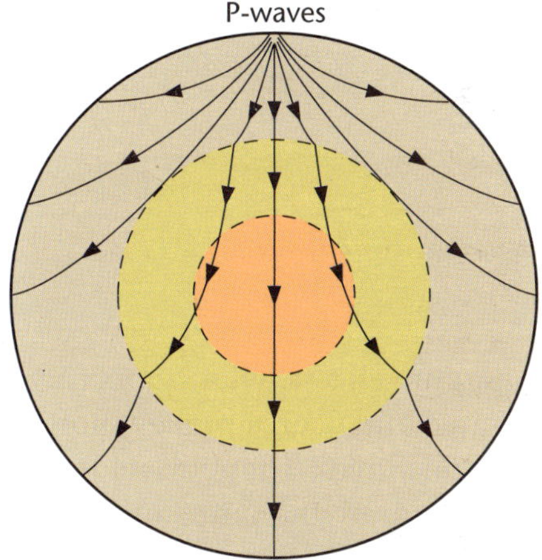

Note: There is only a small change in speed at the boundary between the outer core and inner core. This means that there is only a small change in direction. (It is too small to show on this diagram.)

Using your knowledge

W 3 The diagram shows a P-wave that takes a slightly different path from the P-waves shown on the earlier diagram.

(a) Make a copy of this diagram.

(b) State what is different about the path of this P-wave.

(c) Add to your diagram the P-wave that sets out vertically downwards.

W 4 Do P-waves produce a shadow zone? Explain your answer.

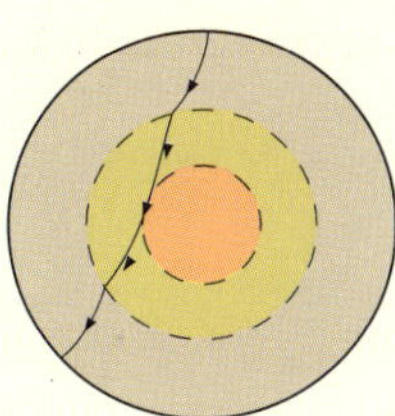

You get a seismograph record of P-waves <u>everywhere</u> on the Earth's surface.

Using ultrasound to detect flaws

To be strong, a metal casting must be solid metal all the way through. There should be no gaps or cracks in it.

Sometimes there are gaps or cracks inside the metal, so you can't see them. You can use ultrasound waves to detect flaws of this kind.

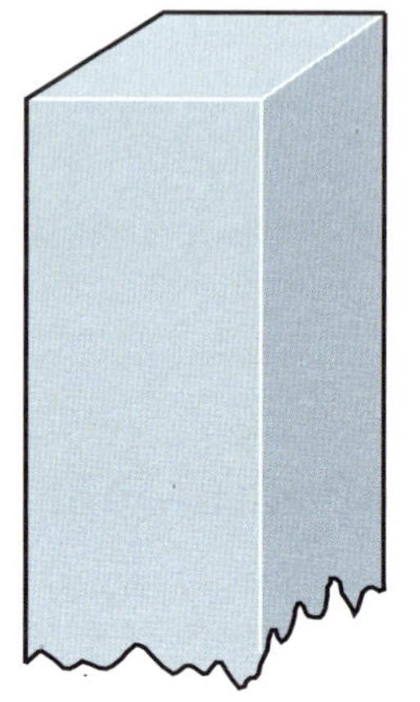

This metal casting needs to be strong. So there should be no gaps or cracks inside it.

The diagram shows what happens with a good casting. The ultrasound source makes short bursts (**pulses**) of ultrasound waves. These travel through the metal and are picked up by the ultrasound detector. The pulses from the detector are processed to make a trace on the screen of a monitor.

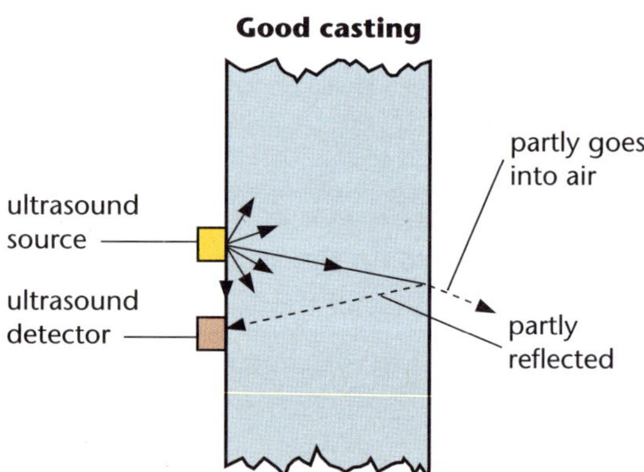

1 **(a)** Make a copy of the picture that you get on the screen from the good casting.

   (b) Explain why the screen shows <u>two</u> pulses.

   (c) Why is the left-hand pulse bigger than the right-hand pulse?

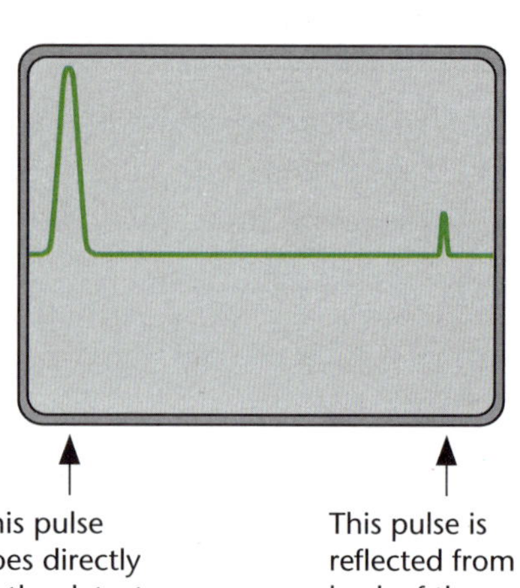

This pulse goes directly to the detector.

This pulse is reflected from the back of the casting.

The diagrams show what happens if there is a crack or gap inside the metal. The ultrasound pulses are also reflected from the front and back of the <u>gap</u>. This means that the detector picks up <u>three</u> different **reflections** of each pulse. So the screen now shows four pulses.

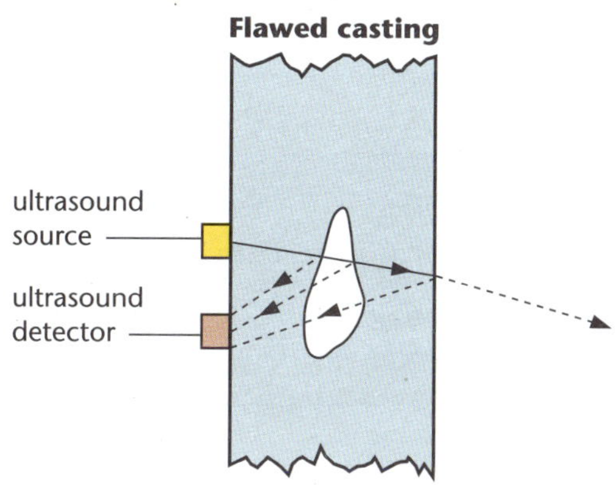

2 **(a)** Make a copy of the picture you get on the screen with the flawed casting. Label the pulses A, B, C and D, as on the diagram.

(b) Underneath your diagram say what each of the pulses A–D represents.

(c) Explain, as fully as you can, why the pulses are progressively smaller.

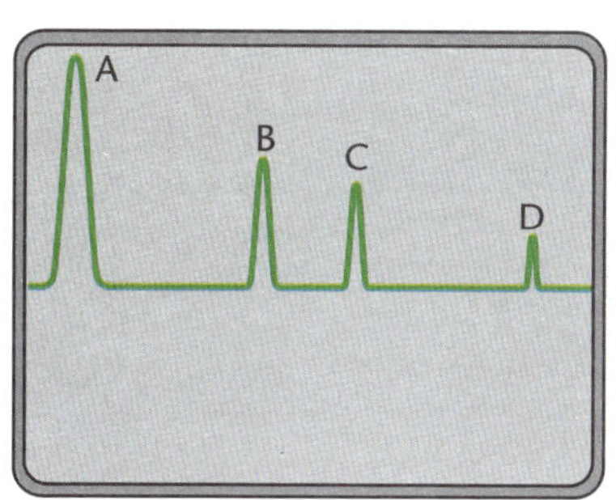

Using your knowledge

W5 The diagram shows the ultrasound trace taken a few centimetres lower down on the casting than the one giving the traces A–D. Describe, in as much detail as you can, what this trace tells you. Give reasons for your answer.

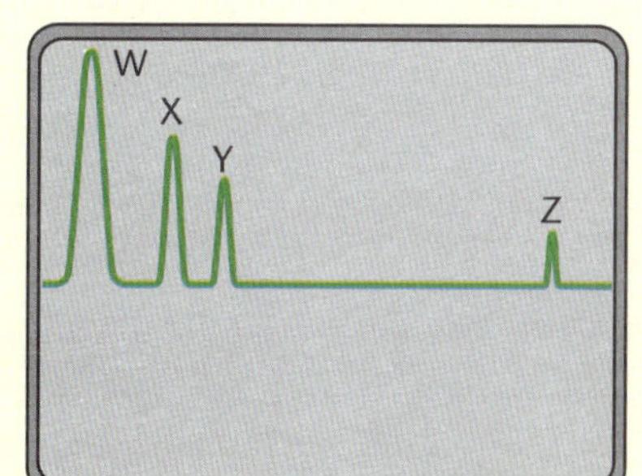

W6 You can also use ultrasound to measure thickness. The reflection from the back of a steel plate occurs 0.1 milliseconds after the pulse is made. How thick is the steel plate?

[The speed of ultrasound in steel is 1500 m/s.]

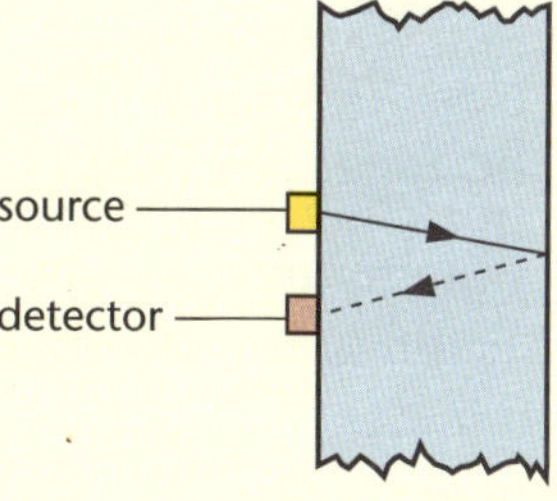

H4

Waves bending round corners

Light travels in straight lines. So when light waves go
past the edge of an opaque object you get a shadow.

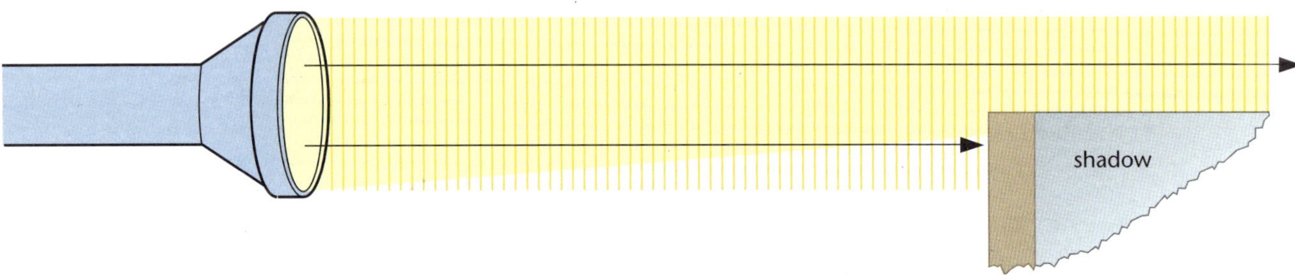

Water waves behave in the same way.
Instead of a shadow that is in darkness,
you get a shadow of still water. But if you
look carefully at the picture, you will see
that the water in the shadow area isn't
completely still. There has been some
bending of the waves around the edge of
the barrier. This is called **diffraction**.

*The edge of the barrier acts as a new source of waves.
These diffracted waves are much smaller than the
original waves. We say that they have a smaller
amplitude.*

1 Explain how the edges of the barrier produce diffracted water waves.

You get diffraction with all types of waves, for example sound waves and radio
waves. The diffraction of sound and radio signals is further evidence that they
do, in fact, travel in the form of waves. But diffraction is much more noticeable
with some sorts of wave than with others.

■ Diffraction of sound waves

Sound waves can be diffracted. This is one of the reasons why you can often hear sounds around the corners of buildings.

The person shown in the diagram can hear the tuba clearly but can hardly hear the flute at all.

2 Copy and complete the sentences.

The sound from the tuba is ____________ round the corner more strongly than the sound from the flute. This is because the sound from the flute has a higher ____________ and a lower ____________.

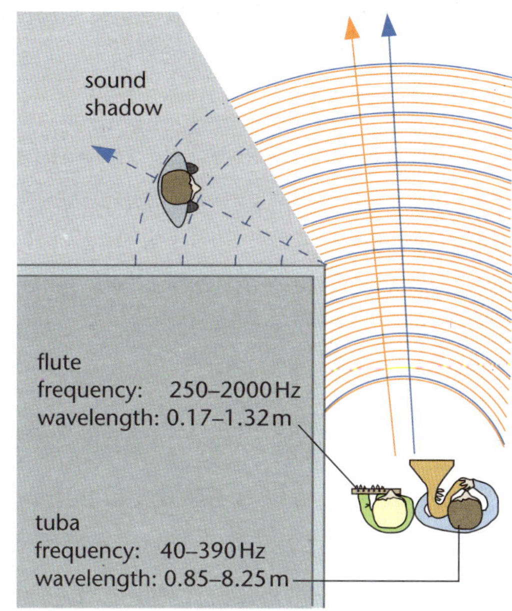

Sound waves are diffracted by the edge of the building, so you can hear the tuba clearly from round the corner.

■ Diffraction of radio waves

Radio waves can also be diffracted. The diagram shows one of the ways that this can happen.

3 Explain how people can often receive radio waves even though they are in the 'radio shadow' of a hill.

As with sound waves, radio waves with longer wavelengths are more strongly diffracted.

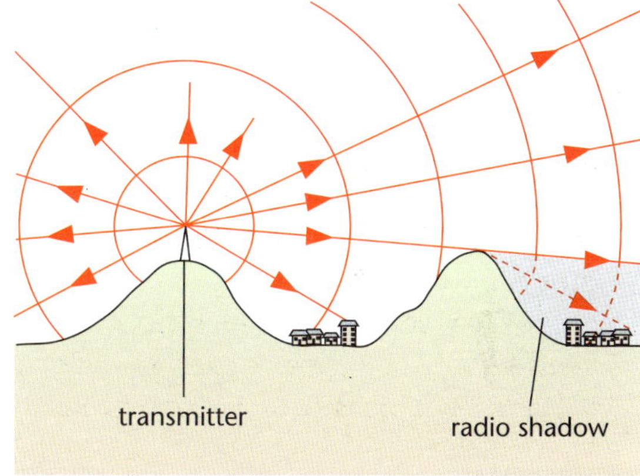

Radio waves are diffracted by the top of the hill. This means that the town receives a signal even though it is in the shadow of the hill.

Using your knowledge

W7 Copy the diagram. Then complete it to show the water waves after they have passed through the gap in the barrier.

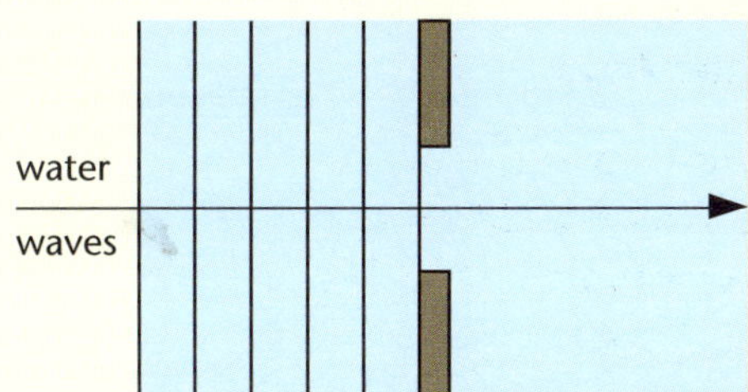

W8 A house in the radio shadow of a hill can obtain VHF (very high frequency) radio signals but not UHF (ultra high frequency) TV signals. Suggest a reason for this.

W9 You get diffraction with all types of electromagnetic waves. But the effects are less noticeable with light waves than with radio waves. Why is this?

H5 A closer look at the radiation from radioactive materials

Follows on from *Science Foundations* Waves and radiation 19–24

Radioactive substances emit three types of **radiation: alpha** (α), **beta** (β) and **gamma** (γ).

Alpha radiation is very easily absorbed, e.g. by thin paper or a few centimetres of air.

Beta radiation is absorbed by a few millimetres of metal.

To absorb gamma radiation you need many centimetres of lead or several metres of concrete.

■ What are alpha, beta and gamma radiation?

The diagrams show some of the differences between α, β and γ radiation.

1 Copy and complete the sentences.

An alpha (α) particle is a ___________ nucleus.
It has a mass of ___________ and an electric charge
of ___________.

A beta (β) particle is an ___________.
It has a mass of ___________ and an electric charge of
___________.

Gamma (γ) radiation is a form of ___________ radiation.
It has a very short ___________.

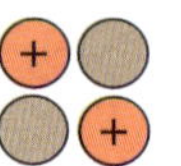
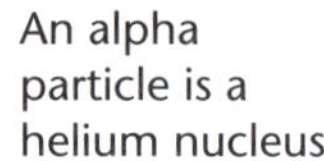

An alpha particle is a helium nucleus.

A beta particle is a fast moving **electron**. It is emitted from the nucleus of an atom.

Symbol	Name	Mass	Charge
+	proton	1	+1
	neutron	1	0
e⁻	electron	almost 0	−1

The electromagnetic spectrum

lowest   increasing frequency   highest

radio waves	micro- waves	infra- red	light	ultra- violet	X-rays	gamma rays

increasing wavelength

longest   shortest

■ Which type of radiation is most dangerous?

Those atoms of elements that are radioactive are called **radio-isotopes** or **radionuclides**. If the radiation from radionuclides is absorbed by living cells, it can kill them or make them cancerous.

The diagrams show which types of radiation are the most dangerous.

2 Answer the following questions, giving a reason for your answer in each case.

(a) Which type of radiation is <u>least</u> harmful when the source of the radiation is <u>outside</u> your body?

(b) Which type of radiation is <u>most</u> harmful when the radionuclide that emits it is <u>inside</u> your body?

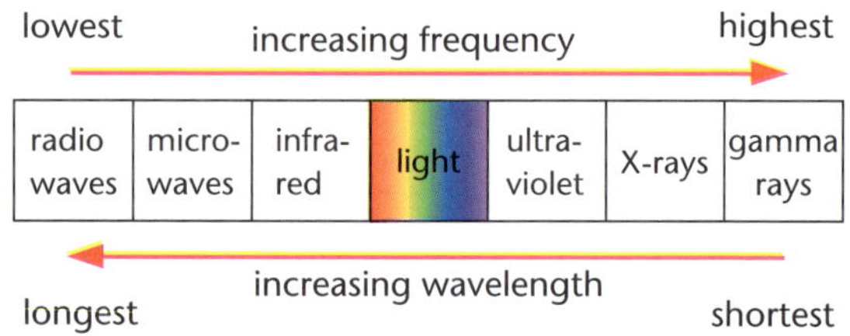

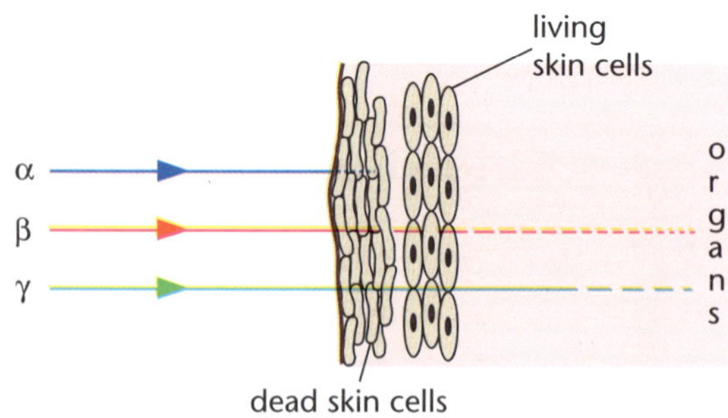

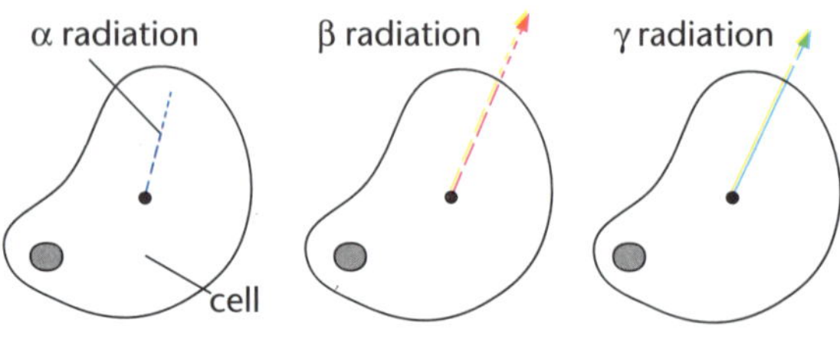

Key ● radionuclide

■ Looking at radioactive decay

You can never tell when a particular radioactive atom
will emit radiation and **decay**. It is a **random** process.
But a sample of radioactive material usually contains
billions of radioactive atoms so, on average, it emits
radiation at a fairly steady rate. Over a period of time,
the sample of radioactive material gradually becomes
less radioactive. The time it takes for the radiation to fall
to half its original level is called the **half-life**. Different
radionuclides have different half-lives. Half-lives can
vary from a tiny fraction of a second to billions of years.

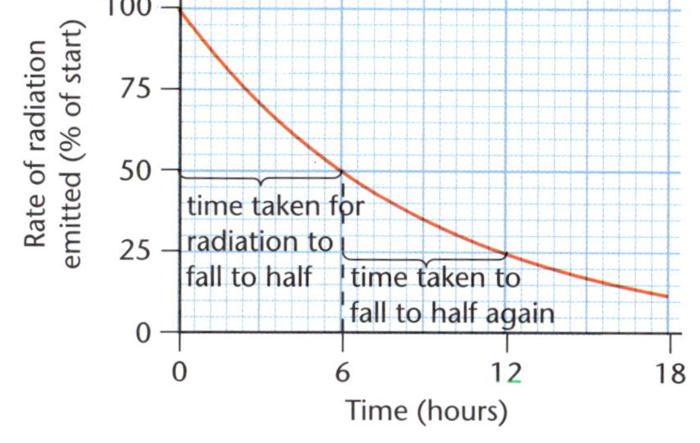

3 **(a)** Make a copy of the graph.

   (b) What is the half-life of this particular radionuclide?

   (c) What proportion of the original radioactive atoms
   in a sample of this radionuclide will have decayed
   after one half-life?

> This line of numbers shows the total number of
> **protons** and **neutrons**. These <u>balance</u>: 226 = 222 + 4.

> This line of numbers shows the number of protons.
> These <u>balance</u>: 88 = 86 + 2.

> **Writing balanced nuclear equations**
>
> A radioactive atom of radium emits an
> alpha particle from its **nucleus** and
> changes into an atom of radon. We can
> write down what happens like this:
>
> $$^{226}_{88}\text{Ra} \rightarrow \,^{222}_{86}\text{Rn} + \,^{4}_{2}\text{He} \;(\alpha\ \text{particle})$$

■ Nuclear equations

When a radioactive atom decays, an atom of a different
element is produced. The box explains how to write
down what happens in the form of a **nuclear equation**.

4 Copy down the following nuclear equations. Under each
one, write down <u>in words</u> what has happened.

   (a) $^{222}_{86}\text{Rn} \rightarrow \,^{218}_{84}\text{Po} + \,^{4}_{2}\text{He}\;(\alpha)$

   (b) $^{228}_{88}\text{Ra} \rightarrow \,^{228}_{89}\text{Ac} + \,^{0}_{-1}\text{e}\;(\beta)$ [Ac = actinium]

> **Example**
>
> An atom of bismuth emits a beta particle
> from its nucleus and changes into an atom
> of polonium. We can write down what
> happens like this:
>
> $$^{214}_{83}\text{Bi} \rightarrow \,^{214}_{84}\text{Po} + \,^{0}_{-1}\text{e}\;(\beta\ \text{particle})$$
>
> Both these lines balance.
>
> The bottom line tells you that a beta
> particle is emitted when a neutron decays
> into a proton.

Using your knowledge

W 10 A radionuclide has a half-life of 100
years.

   (a) How long will it take a sample of the
   radionuclide to become $\frac{1}{4}$ as
   radioactive?

   (b) What proportion of the radioactive
   atoms have decayed when the sample
   is 10% as radioactive as at the start?

W 11 A radionuclide becomes one-third
as radioactive every 10 minutes.
What is its half-life?
[Hint: plot a graph.]

Using radioactive materials

Follows on from *Science Foundations* **Waves and radiation 22–24 and** *Extension* **Waves and radiation H5**

The half-life of a radionuclide is the time it takes for half of the radioactive atoms to decay.

■ Dating rocks

Uranium atoms are radioactive. They have a very long half-life. They decay to produce a series of radionuclides, each of which has a relatively short half-life. When the last of this series of radionuclides decays, it produces stable atoms of lead.

Some igneous rocks contained uranium atoms when they were first formed. We can date these rocks by comparing the numbers of uranium atoms and lead atoms that are now in the rock. The example explains how we can do this.

1 (a) Make a copy of the graph.

(b) An igneous rock contains 1 atom of uranium-238 for every 3 atoms of lead-206 (formed from the decay of the uranium). How old is the rock? [Show all of your reasoning.]

Other rocks contained atoms of a potassium radio-isotope when they were first formed. These atoms decay to form a stable isotope of a gas called argon. Sometimes this argon gets trapped inside the rock. We can then date the rock by comparing the numbers of potassium-40 and argon-40 atoms that it now contains.

2 A rock has three times as many argon-40 atoms as potassium-40 atoms. Use the information in the box to work out the age of the rock. [Show all of your reasoning.]

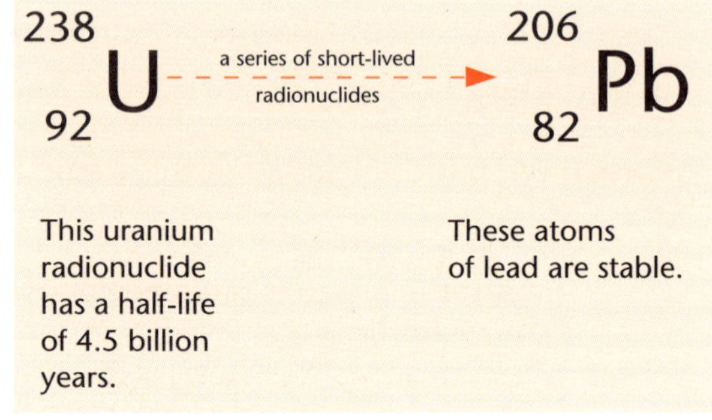

This uranium radionuclide has a half-life of 4.5 billion years.

These atoms of lead are stable.

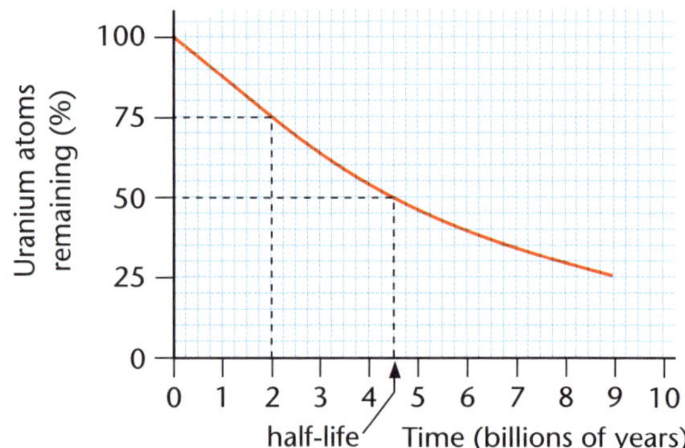

Example
A sample of rock contains 3 atoms of uranium-238 for every 1 atom of lead-206. In other words, 75% of the uranium-238 atoms are still there. From the graph, the rock is about 2 billion years old.

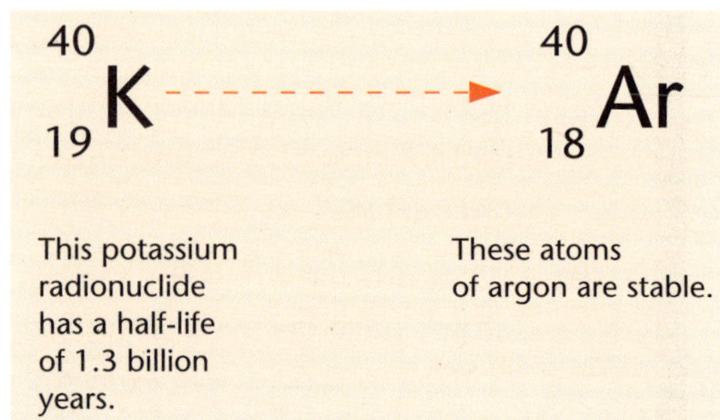

This potassium radionuclide has a half-life of 1.3 billion years.

These atoms of argon are stable.

Using radioactive tracers

Small amounts of radionuclides can be used to find out what happens to substances inside the bodies of plants and animals. The picture shows an example. Radionuclides that are used in this way are called **tracers**.

3 (a) What would be a suitable half-life for the iodine radionuclide used as a tracer – seconds, minutes, hours, days or weeks?

(b) What type of radiation should the iodine radionuclide emit? Give <u>two</u> reasons for your answer.

A doctor injects her patient with a radionuclide of iodine.

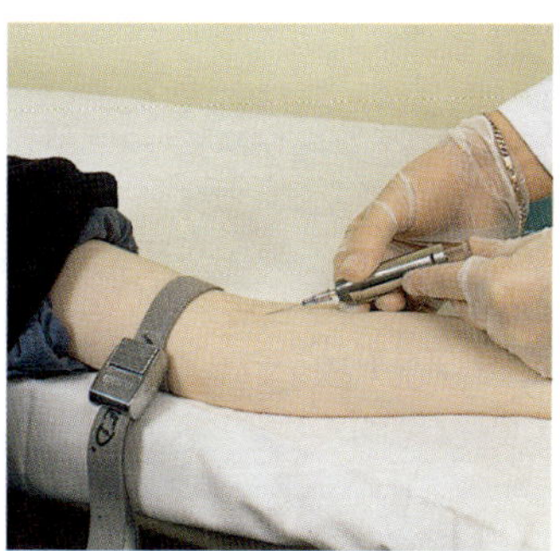

After a few hours, she can measure the uptake of iodine by the patient's thyroid gland. She does this with a radiation detector <u>outside</u> the patient's body.

> **REMEMBER**
> **Alpha** radiation is very easily absorbed. Inside your body it can cause serious damage to your cells.
> **Beta** radiation can pass quite easily through the soft parts of your body; **gamma** radiation can pass through very easily indeed.

Nuclear fission

In nuclear reactors, atoms with a very large **nucleus** – for example, uranium atoms – are bombarded with neutrons. The diagram shows what happens.

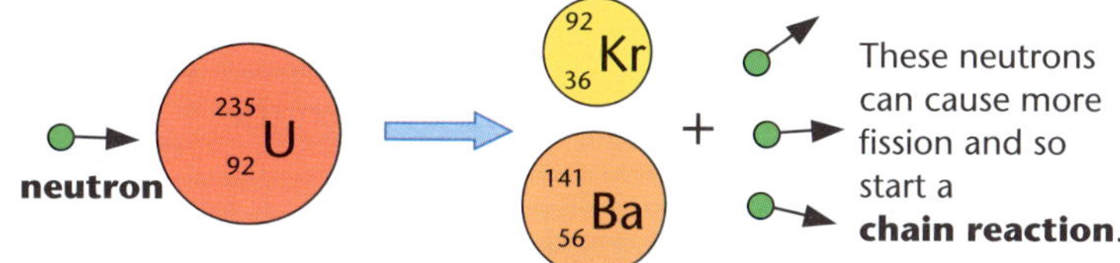

*The **fission** (splitting) of a uranium atom.*

The new atoms produced by fission are radioactive.

4 Describe <u>in words</u> the nuclear fission reaction shown in the diagram. [Your answer should include why the reaction is called nuclear fission.]

5 The fission of one gram of nuclear fuel can release about a million times more energy than burning one gram of an ordinary fuel such as coal or gas. Explain why.

> **Did you know?**
> The fission of an atom releases a lot more energy than when two atoms make a chemical bond.

Using your knowledge

W 12 (a) Draw a graph showing the decay of potassium-40 over three half-lives.

(b) Use the graph to estimate the age of a rock that contains two potassium-40 atoms for every three argon-40 atoms.

W 13 Scientists suspect that chemical pollution in a river is coming from a waste dump.

(a) How could they use a radionuclide to find out?

(b) What properties should the radionuclide have to make this test as effective and as safe as possible?

Pressure in gases

You can calculate a **pressure** using $\quad\text{pressure} = \dfrac{\text{force}}{\text{area}}$

We live at the bottom of deep layer of air called the atmosphere.
The weight of the atmosphere above us produces a large pressure.

If you increase the pressure on some gas, you can squeeze it into a smaller space. To understand exactly what is happening, you need to <u>measure</u> the pressure and the volume of the gas. The diagrams show some of these measurements.

1 (a) Copy and complete the table for the three sets of measurements shown on the diagrams.

Pressure of gas (N/cm^2)	Volume of gas (cm^3)	Pressure × volume

(b) What effect does doubling the pressure have on the volume of the gas?

(c) What do you notice about the (pressure × volume) of the gas in all three cases?

The standard units for measuring pressure are N/m^2 or Pa (**pascals**). But you don't always have to use these. In the table above, N/cm^2 was a more convenient unit to use. We could also have used atmospheric pressure as the unit: normal atmospheric pressure is called **1 atmosphere**.

2 Make another copy of the table. This time use atmospheres as your unit of pressure.

If you take measurements for lots of different pressures, you always get the same figure for (pressure × volume), provided that no gas escapes and the temperature stays the same.

A very precise way of stating the relationship between the pressure and the volume of some gas is shown in the box.

> **Pressure and volume of a gas**
> For a fixed mass of gas at constant temperature, the volume of the gas is inversely proportional to the pressure.

The following formula connects the pressure and volume of a gas:

$$\text{initial pressure} \times \text{initial volume} = \text{final pressure} \times \text{final volume}$$

3 The syringe of gas shown at the top of the opposite page is taken to the top of a mountain where the temperature is the same but the atmospheric pressure is only $8\,\text{N/cm}^2$. Calculate the volume of the gas. [Start by writing down the formula. Show your working.]

> **Example**
>
> With a 3 kg mass on the syringe of air shown opposite, the pressure is $16\,\text{N/cm}^2$.
>
> $$\text{initial pressure} \times \text{initial volume} = \text{final pressure} \times \text{final volume}$$
>
> $$10 \times 60 = 16 \times \text{final volume}$$
>
> $$\text{so final volume} = \frac{10 \times 60}{16}$$
>
> $$= \underline{37.5\,\text{cm}^3}$$

■ Explaining gas pressure

A gas has a pressure because its molecules constantly bombard anything that the gas surrounds or that the gas is contained inside.

If some gas inside a container is squeezed into <u>half</u> the space, its molecules will collide with the sides of the container <u>twice</u> as often. So the pressure will be <u>twice</u> as big.

The pressure will be exactly twice as big only if the temperature of the gas stays the same. If the gas becomes hotter, as it often does because of the work you do (energy you transfer) as you force it into a smaller space, the particles move about faster. This makes the pressure a little higher.

4 You squeeze some gas into one-tenth of its original volume. Then you wait a while to let it cool down to the same temperature.

   (a) What happens to the pressure of the gas?

   (b) Explain your answer in terms of particles.

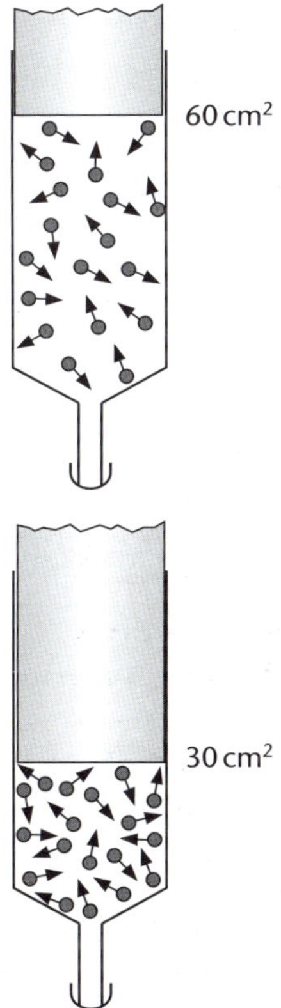

> ## Using your knowledge
>
> **F1 (a)** 100 litres of air is forced into a 35 litre steel cylinder. How many times bigger than atmospheric pressure would you expect the pressure of the air to be?
>
> **(b)** When you actually measure the pressure inside the cylinder, you find that it is higher than you expected. Suggest an explanation for this.
>
> **F2** Why does the pressure of a gas decrease when you increase its volume?

More about hydraulic systems

Follows on from *Science Foundations* Forces 8–10

You can calculate a pressure using $\text{pressure} = \dfrac{\text{force}}{\text{area}}$

The pressure in a liquid acts equally in all directions.
Hydraulic systems use liquids to make forces act in the right places and in the right directions.

Hydraulic systems contain a liquid that is fully enclosed. When the liquid is put under pressure, there is the <u>same</u> pressure everywhere in the liquid. So we can use the hydraulic system to make a small force on a small area balance a larger force on a larger area. The diagram shows an example of this.

1 Explain how the hedgehog can balance the pig.

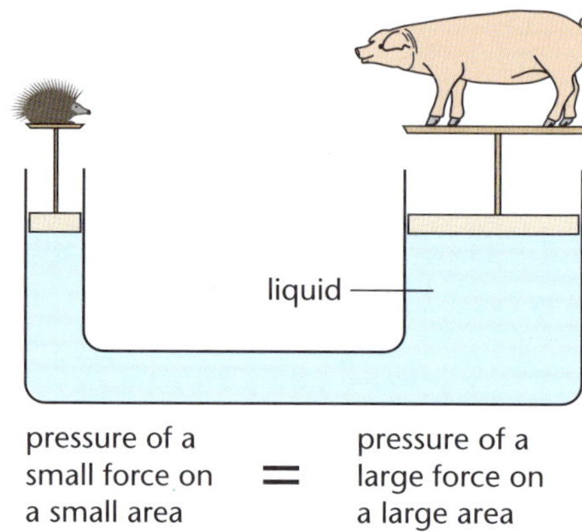

We often use hydraulic systems to produce larger forces than the ones we apply to them. We use them as **force multipliers**.

Pressure is given by $\text{pressure} = \dfrac{\text{force}}{\text{area}}$, so

force = pressure × area

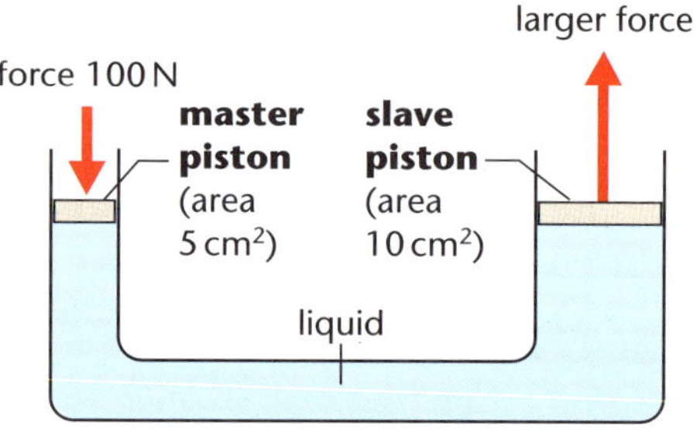

In the hydraulic system shown in the diagram, the slave piston has <u>double the area</u> of the master piston. This means that the slave piston produces <u>double the force</u>:

$2 \times 100\,\text{N} = \underline{200\,\text{N}}$

Sometimes you need to know the actual pressure in the hydraulic liquid. The box shows how you can calculate this.

> **Example**
> At the master piston:
> $$\text{pressure} = \frac{\text{force}}{\text{area}}$$
> $$= \frac{100\,\text{N}}{5\,\text{cm}^2}$$
> $$= \underline{20\,\text{N/cm}^2}$$

There is the same pressure all through the liquid.
So at the slave piston:

$$\begin{aligned}
\text{force} &= \text{pressure} \times \text{area} \\
&= 20 \times 10 \\
&= \underline{200\,\text{N}}
\end{aligned}$$

[This is the same answer as we got before, but we have calculated it in a different way.]

> **Note**
> Because of the weight of the liquid itself in a hydraulic system, the pressure varies with depth. But, for small differences in depth, the differences in pressure are quite small, so we usually ignore them.

2 For the system shown on the right:

(a) Calculate the pressure of the master piston on the liquid.
[Start by writing down the formula. Show all your working.]

(b) Calculate the lifting force produced by the slave piston. Give your reasoning.

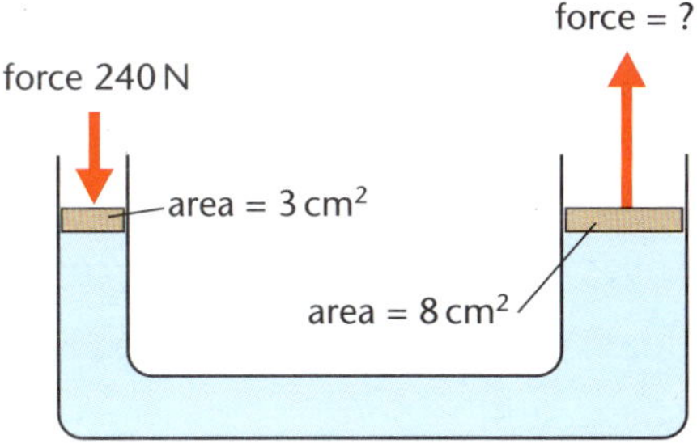

■ Some other advantages of hydraulic systems

Hydraulic systems don't just multiply forces. They are also very useful for sending forces round awkward corners to where they are needed and for making the forces act in the right direction when they get there.

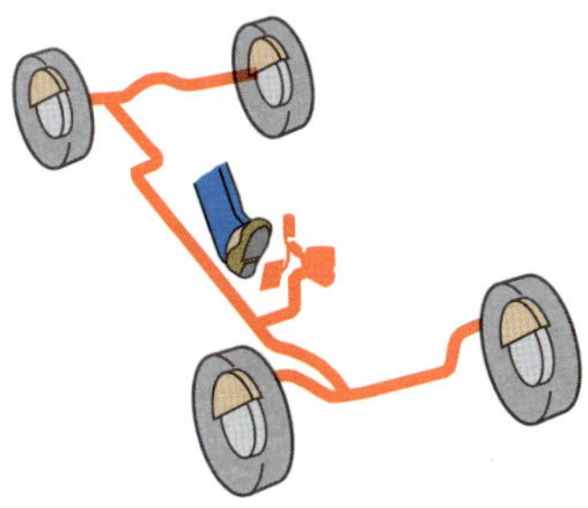

With hydraulic brakes, it is easy to send exactly the same force to each wheel of a car. This means that the car doesn't swerve when you use the brakes.

The diagrams show some other advantages of using hydraulic systems.

3 State <u>four</u> reasons why cars use hydraulic braking systems.

4 What <u>two further</u> reasons make hydraulic systems suitable for aircraft control surfaces?

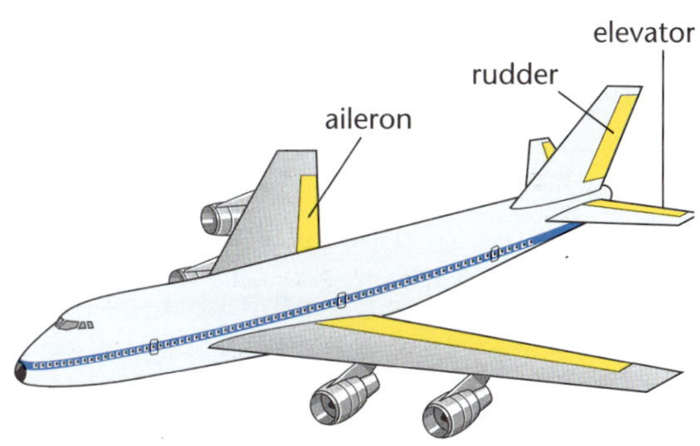

Hydraulic systems can send large forces to the control surfaces, but they don't take up very much space.

Using your knowledge

F3 **(a)** In a hydraulic system, the area of the master piston is 1.5 cm² and the area of the slave piston is 75 cm².
What force on the master piston will produce a slave piston force of 2500 N?

(b) What will the pressure in the hydraulic fluid be when applying this force?

H3

More about orbits

Follows on from *Science Foundations* **Forces 14–15**
The force of **gravity** between masses keeps satellites in their orbits around planets, and planets in their orbits around the Sun.

■ Orbits of satellites

To stay in its **orbit** at a particular distance, **a satellite** must move at a particular speed around a planet.

The orbit that a satellite is put into depends on the job that you want the satellite to do. The diagram shows the two main types of orbit that are used.

1 Explain, as fully as you can:

  (a) why a monitoring satellite such as a weather satellite is usually put into a polar orbit;

  (b) why a communications satellite is usually put into a geostationary orbit.

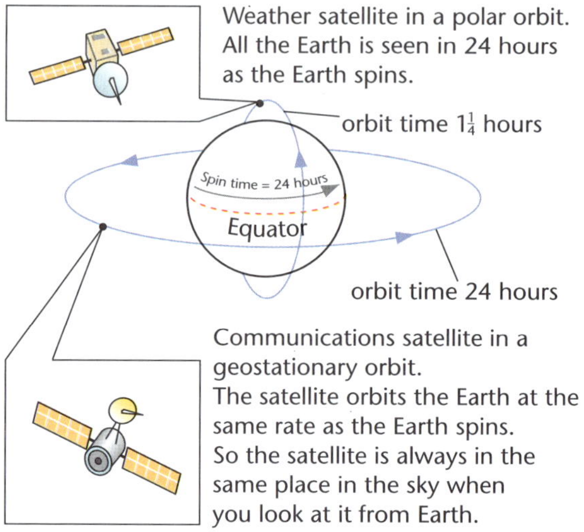

■ Orbits of planets

To stay in its orbit at a particular distance, a **planet** must move at a particular speed around the **Sun**.

The table shows the orbit **periods** of the planets and their average distances from the Sun.

2 (a) What happens to the orbit period when the distance of a planet from the Sun doubles?
[Hint: compare Saturn and Uranus.]

  (b) How many times further from the Sun does a planet have to be for its orbit period to double?
[Hint: compare Earth and Mars.]

3 (a) Use a calculator to work out the values of $\dfrac{[\text{period}]^2}{[\text{distance}]^3}$ for each planet.

  (b) What do you notice about the values?

Planet	Average distance from Sun (Earth = 1)	Time taken for one orbit (Earth = 1)
Mercury	0.4	0.24
Venus	0.7	0.62
Earth	1.0	1.0
Mars	1.5	1.9
Jupiter	5.2	11.9
Saturn	9.6	29.5
Uranus	19.2	84.1
Neptune	30.1	165
Pluto	39.5	249

Kepler's third law
The relationship between the period of a planet and its distance from the Sun is quite complicated:

$$[\text{period}]^2 \propto [\text{distance}]^3$$

($\propto$ means 'is proportional to'.)

This relationship was discovered by the astronomer Johannes Kepler nearly 400 years ago.

We have been using the <u>average</u> distances of planets from the Sun because the orbits of planets are not, in fact, circles. They are an oval shape called an **ellipse**.

The orbits of most of the planets are, however, quite close to being circular. They are only slightly elliptical.

4 Which planet has a much more elliptical orbit than the others?

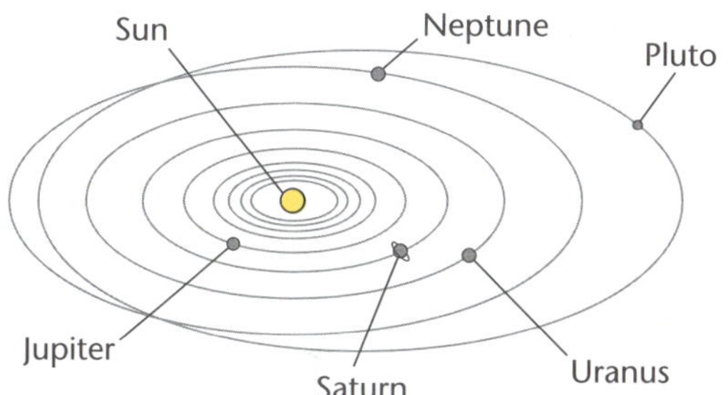

■ Orbits of comets

Comets are balls of ice and rock, usually a few kilometres in diameter. The diagrams show what the orbit of a comet is like and when you can see a comet.

5 (a) Describe the orbit of a comet.

(b) Explain <u>when</u> you can see a comet and <u>why</u> you can see it.

Comet West.

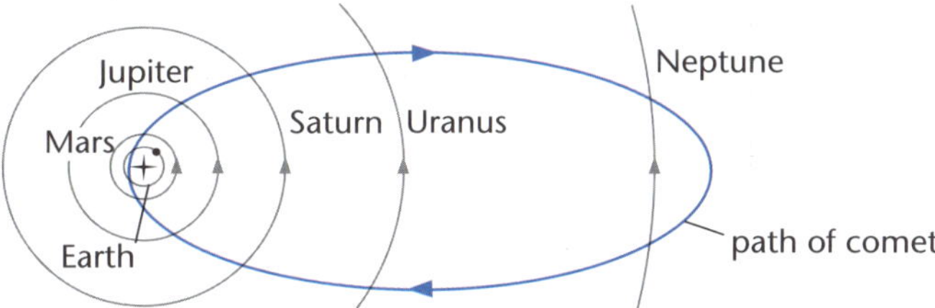

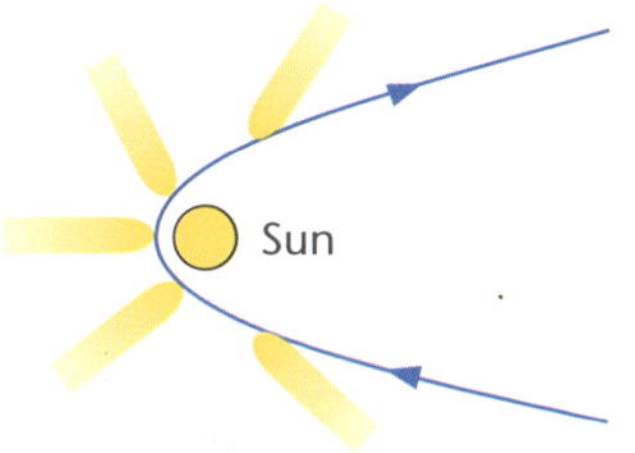

Comets have very large, very elliptical orbits. They can take tens, or even thousands, of years to make one orbit. During the most distant parts of their orbits, comets move more slowly.

You can see comets when they are closest to the Sun. Energy from the Sun melts the ice, and the 'solar wind' makes the vapour into a tail.

Using your knowledge

F4 The Moon has no atmosphere. An astronaut is at the top of a mountain on the Moon. She fires bullets from three different guns. The diagram shows what happens to the bullets.

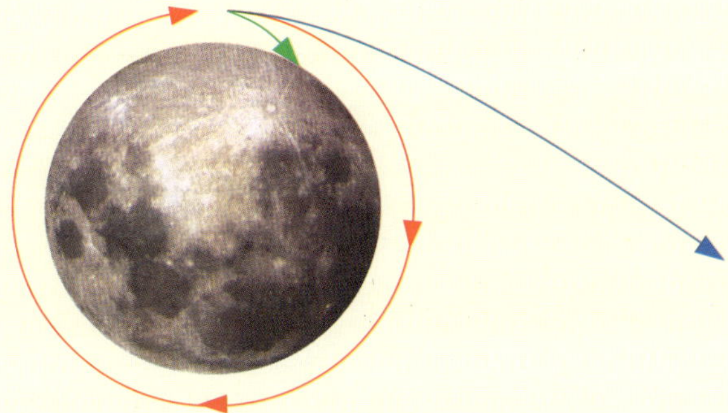

(a) Make a copy of the diagram and add the following labels:

- slow moving bullet
- faster moving bullet
- very fast moving bullet
- escapes into space
- pulled by gravity to the surface
- goes into orbit around the Moon

(b) It would <u>not</u> be possible to make a bullet go into orbit around the Earth by firing it at just the right speed from the top of Mount Everest. Why not?

H4

The life history of a star

Follows on from *Science Foundations* Forces 16

Stars form when thinly spread out matter is pulled together by the force of **gravity**.

Billions of years ago, this spread out matter was mainly hydrogen and helium.

In stars like the Sun, hydrogen atoms join together to form helium.

This process is called **nuclear fusion**; it releases a lot of energy and causes a small loss in mass.

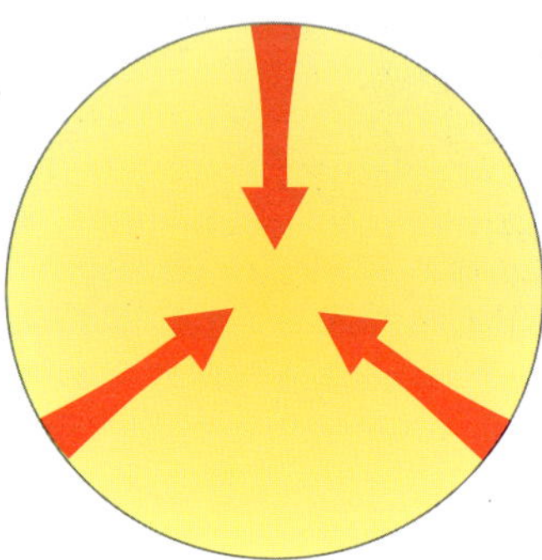

A star has a very big mass, so the force of its own gravity tends to make it collapse. This makes the core of the star hot enough for nuclear fusion to occur. The bigger the mass of the star, the faster these reactions are.

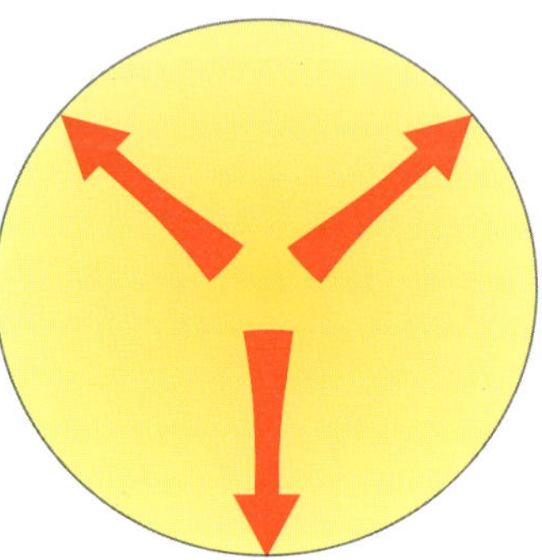

The energy released by the nuclear fusion reactions produces a **pressure** which tends to make the star expand.

The **Sun** is an average star that is in the stable period of its life. It has been more or less like it is today for the past 5 billion years, and it will stay the same for another 5 billion years. The diagrams show why.

When most of the Sun's hydrogen has been converted into helium, the Sun will expand into a **red giant**. This will be about ten times the present diameter of the Sun. The inner planets will be vaporised. Only the rocky cores of the outer planets will remain.

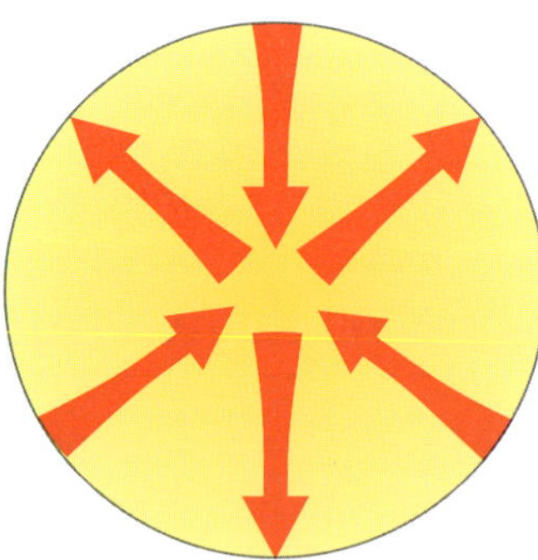

During the main period of a star's life, the gravitational forces and the outward pressure forces are balanced. So the star is **stable**.

Part of the red giant that will eventually form from the star shown above.

Different nuclear fusion reactions in a red giant produce bigger and bigger atoms up to the size of iron atoms.
Fusion reactions for making atoms that are bigger than iron atoms do not release energy. So they don't occur in a red giant.

What happens after the red giant stage depends on the
mass of the star.

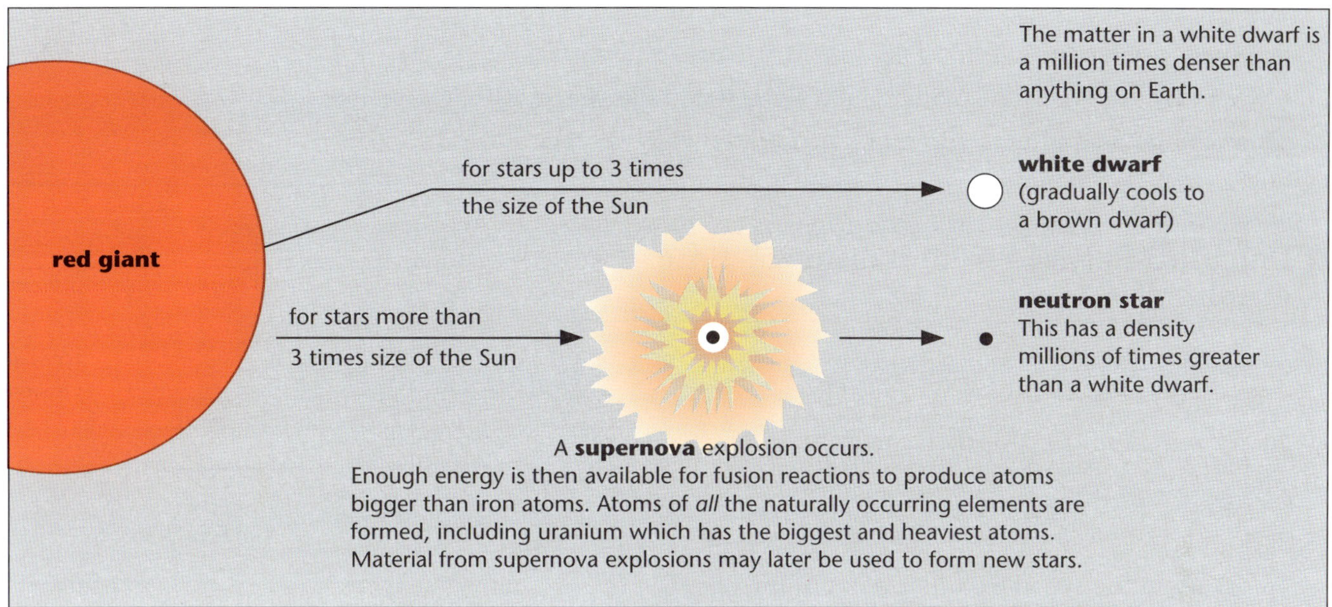

1   Why is the Sun at present in the middle
of a very long stable period?

2   Describe, as fully as you can, what will
happen to the Sun starting from about
5 billion years in the future.

3   What would be different about the Sun's
future if it were about four times as
massive as it actually is?

4   Where did the material come from that
formed the Sun and the planets?
Give reasons for your answer.

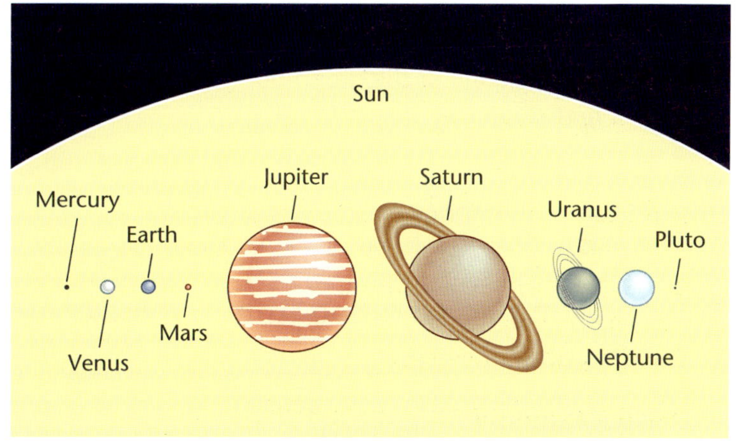

*The solar system contains all of the naturally occurring elements,
up to and including uranium. So some of the material from which
it was formed must have come from a supernova explosion.*

Using your knowledge

F 5   The table shows how long stars of
different masses exist as stable stars like
the Sun.

(a)   What pattern do these figures show?

(b)   Can you suggest a possible reason for
the pattern?

Mass of star (Sun = 1)	Life as a star like the Sun (billions of years)
3.0	0.5
1.0	10
0.5	200

More about speed and acceleration

> **Follows on from** *Science Foundations* Forces 5 and 19
>
> A steep slope on a **distance–time** graph indicates a high **speed**.
>
> A steep slope on a **velocity–time** graph indicates a high **acceleration**.

■ Calculating speed from a distance–time graph

$$\text{Speed} = \frac{\text{distance}}{\text{time}}$$

So, for part A of the graph on the right:

$$\text{speed} = \frac{20}{5} = \underline{4 \text{ m/s}}$$

To calculate the speed for other parts of the graph, you need to use the slope or **gradient** of the graph. The example shows you how to do this for part B of the graph.

1 (a) Make a copy of the graph but mark the changes for part C instead of part B.

(b) Calculate the speed for part C of the graph. [Show your working.]

■ Calculating acceleration from a velocity–time graph

$$\text{Acceleration} = \frac{\text{change in velocity}}{\text{time taken}}$$

So for part P of the graph on the right:

$$\text{acceleration} = \frac{2}{2} = \underline{1 \text{ m/s}^2}$$

To calculate the acceleration for other parts of the graph, you need to use the slope or gradient of the graph. The example shows you how to do this for part Q of the graph.

2 (a) Make a copy of the graph but mark the changes for part R instead of part Q.

(b) Calculate the acceleration for part R of the graph. [Show your working.]

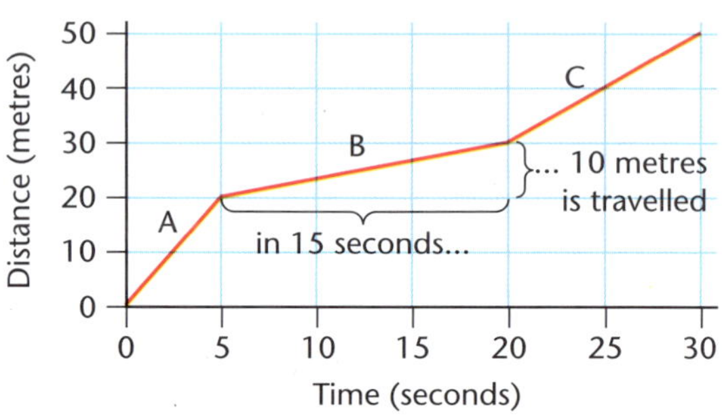

> **Example**
> In part B of the graph:
>
> $$\text{speed} = \frac{\text{distance}}{\text{time}} = \frac{10}{15} = \underline{0.67 \text{ m/s}}$$

> **A note about speed and velocity**
> The **velocity** of an object includes the <u>direction</u> that it is travelling as well as its <u>speed</u>.
>
> Velocity–time graphs, however, are about objects moving in the <u>same</u> direction along a <u>straight line</u>. This means that you don't have to worry about the difference between speed and velocity. You can think of velocity as just another word for speed.

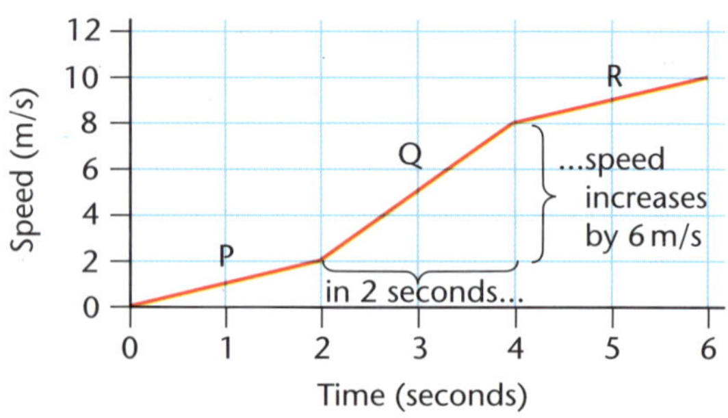

> **Example**
> In part Q of the graph:
>
> $$\text{acceleration} = \frac{\text{change in velocity}}{\text{time}} = \frac{6}{2} = \underline{3 \text{ m/s}^2}$$

What if the graphs are curved?

Distance–time graphs are not always made up of straight lines. The graph on the right, for example, is curved. The line on the graph is getting less and less steep. This tells you that the object is slowing down.

3 **(a)** Make a copy of the graph.

(b) Copy and complete the sentences.

The slope (or gradient) of the graph becomes gradually less ___________. This means that the object is moving more and more ___________.

Velocity–time graphs can also be curved. This means that the acceleration is changing.

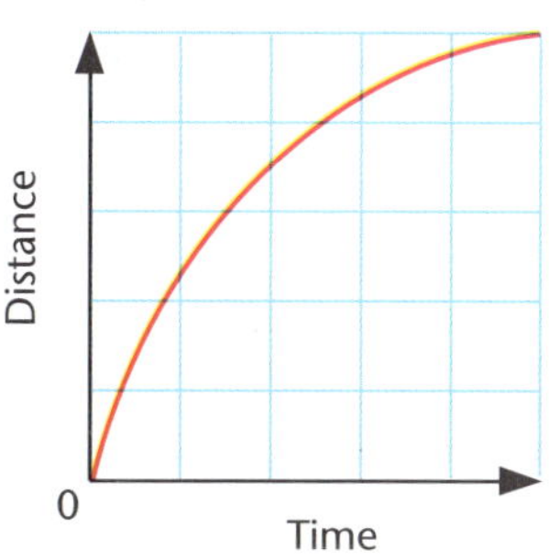

Using area to calculate distance
Area below graph = average height × width
But average height = average speed
and   width = time
So
  area below graph = average speed × time
                   = distance travelled

Calculating distance travelled from a velocity–time graph

The area beneath a velocity–time graph tells you the distance that an object has travelled. The box explains why.

The example shows how you can work out from the graph on the right the distance travelled in the first five seconds.

4 Make a copy of the graph, but mark the area under the graph for between 5 and 10 seconds. Then calculate the distance travelled during this period. [Show your working.]

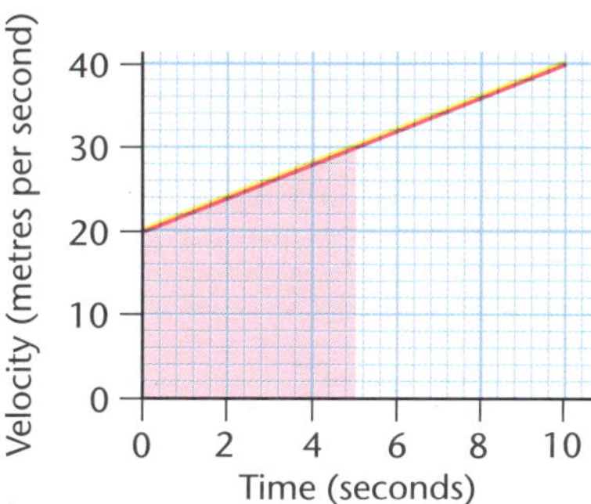

Example
For 0–5 seconds:

$$\text{area below graph} = \frac{20 + 30}{2} \times 5 = \underline{125 \text{ metres}}$$

Using your knowledge

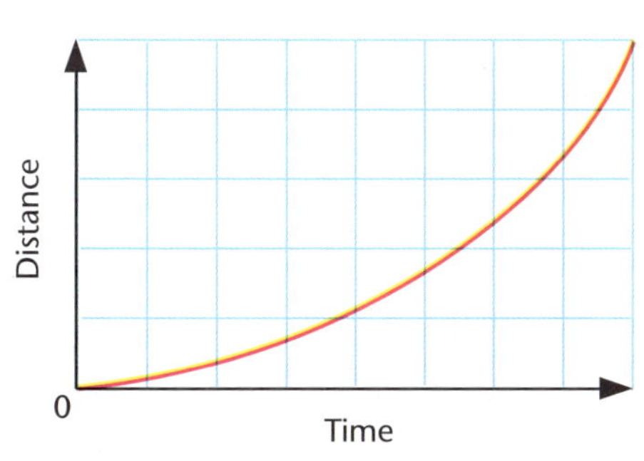

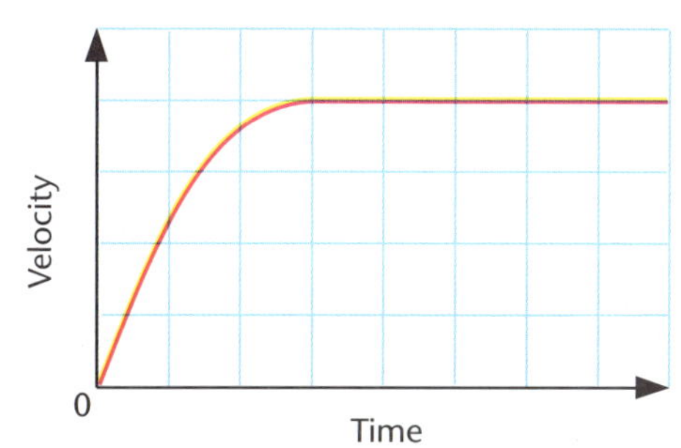

F6 What is happening to the gradient on the graph?

What is happening to the speed?

F7 What is happening to the gradient on the graph?

What is happening to the acceleration?

Looking at kinetic energy

Follows on from *Science Foundations* **Forces 23**

The energy an object has because it is moving is called **kinetic energy**.

The bigger the mass of an object and the faster it is moving, the more kinetic energy it has.

You can work out the kinetic energy of a moving object using this formula:

$$\text{kinetic energy} = \tfrac{1}{2} \times \text{mass} \times [\text{speed}]^2$$
$$\text{(joules, J)} \qquad \text{(kilograms, kg)} \qquad [\text{(m/s)}]^2$$

The example shows how to work out the kinetic energy of a ball that is thrown up into the air.

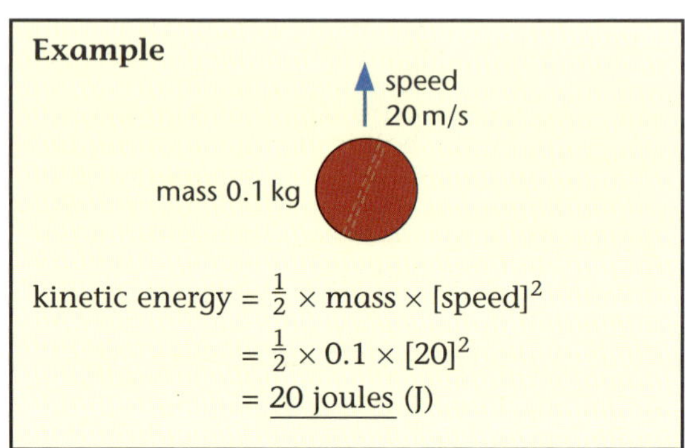

$$\text{kinetic energy} = \tfrac{1}{2} \times \text{mass} \times [\text{speed}]^2$$
$$= \tfrac{1}{2} \times 0.1 \times [20]^2$$
$$= \underline{20 \text{ joules (J)}}$$

1 A car of mass 750 kg is moving at a speed of 20 metres per second. Calculate its kinetic energy.
[Start by writing out the formula. Show your working.]

■ Why fast-moving things can do a lot of damage

In most collisions, the kinetic energy of the objects that collide is transferred to their internal structure. This can bend or break the objects. It also usually makes them warmer.

The more kinetic energy a moving object has, the more damage it can do in a collision. It can do this damage to itself, to whatever it collides with, or to both.
An example of this is explained in the box.

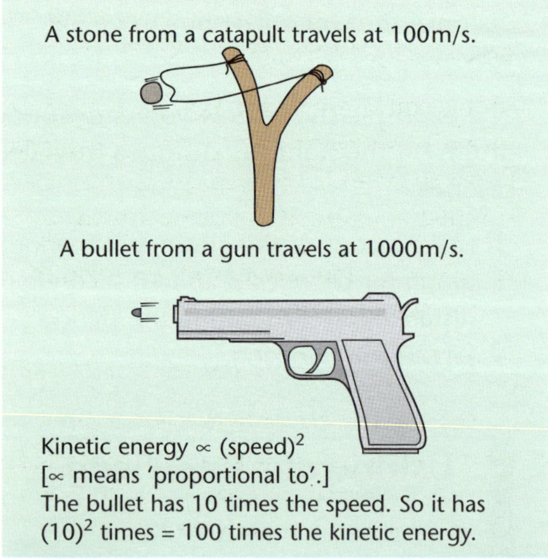

Kinetic energy $\propto$ (speed)^2
[$\propto$ means 'proportional to'.]
The bullet has 10 times the speed. So it has
$(10)^2$ times = 100 times the kinetic energy.

2 The speed limit in some residential streets is now 20 miles per hour rather than 30 miles per hour. How many times less kinetic energy has a car travelling at 20 mph than it has at 30 mph?

What goes up usually comes down

If you throw a ball straight up into the air, it gradually
slows down, stops, and then falls back to the ground.
The diagram shows what is happening to the energy of
the ball as it does this.

3 Copy and complete the table.

	Type of energy the ball has (kinetic or potential)
When you first throw the ball up from the ground	
When the ball is at its highest point	
When the ball reaches the ground again	
When the ball is halfway up (or down)	

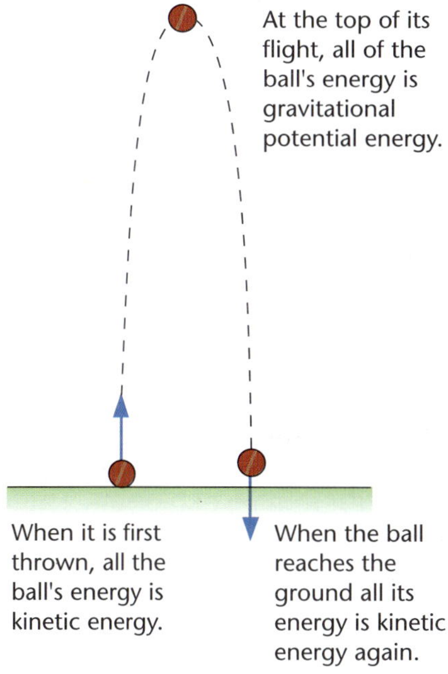

Escaping from the Earth

If you could throw the ball fast enough, it would <u>never</u>
lose all its kinetic energy. So it would escape from the
Earth altogether. The speed that is needed for this to
happen is called the **escape speed**.

4 Use the information from the diagram to answer the
following questions.

(a) What is the escape speed from the Earth's surface?

(b) Quite a lot of hydrogen escapes into the Earth's
atmosphere from industrial processes. But the
atmosphere contains hardly any hydrogen at all.
Suggest a reason for this.

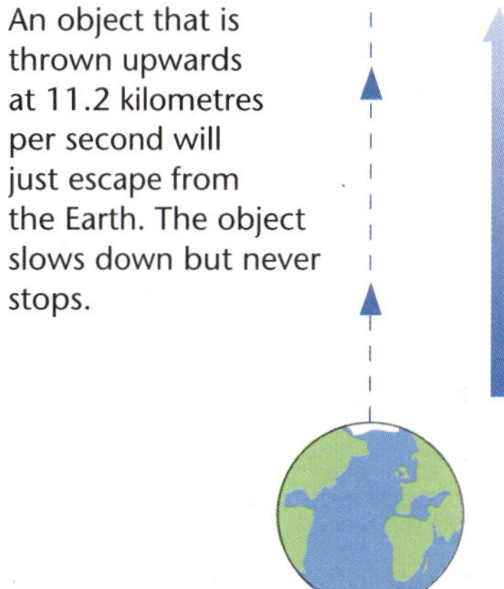

How fast do gas molecules move?
The average speed of hydrogen
molecules in the Earth's atmosphere
is greater than 11.2 km/s.
The average speed of oxygen
and nitrogen molecules in the
Earth's atmosphere is less than 11.2 km/s.

Using your knowledge

F8 The chance of being killed in a road
accident is related to the kinetic energy
of the moving vehicles. How many times
more kinetic energy does a car have
when its speed increases from 50 miles
per hour to 70 miles per hour?

F9 The escape speed from the surface of the
Moon is 2.4 kilometres per second.
Suggest a reason why there is no
atmosphere at all on the Moon.

How did the Universe begin?

Follows on from *Science Extension* **Forces H6**

During the 1920s, the astronomer Edwin Hubble discovered that distant **galaxies** are moving away from us. He also discovered that the further away galaxies are from us, the faster they are moving away.

This suggests that, at one time, all the matter in the **Universe** was in the same place and that the Universe began with a huge explosion. This explosion is called the '**big bang**'.

1 (a) Explain what is meant by the 'big bang' theory.

(b) Why do astronomers think that the Universe began in this way?

By measuring the speeds of many galaxies and their distances from us, astronomers have calculated <u>when</u> the 'big bang' happened. They think that it was probably between 15 and 18 billion years ago.

Astronomers measure the huge distances to stars and galaxies in light-years. A **light-year** is the distance that light travels in a year. Light travels 300 million metres every <u>second</u>, so it travels a very long way in a whole year.

A galaxy that is 1.6 billion light-years away is travelling away from us at about one-tenth of the speed of light.

$$\text{speed} = \frac{\text{distance}}{\text{time}}$$

So

$$\text{time} = \frac{\text{distance}}{\text{speed}}$$

$$= \frac{1.6 \text{ billion}}{0.1}$$

$$= 16 \text{ billion years}$$

The 'big bang' theory of the beginning of the Universe

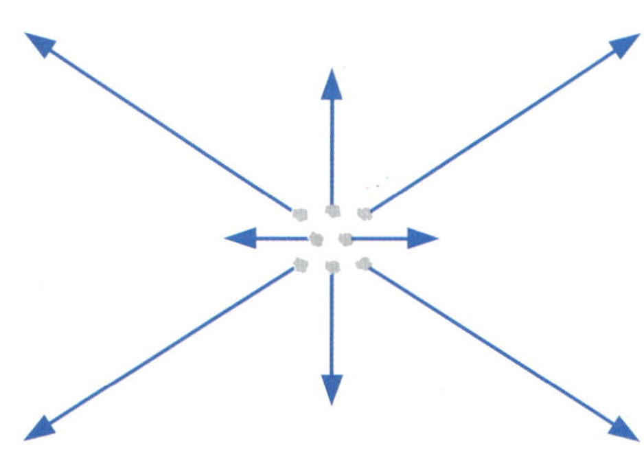

To begin with, all the matter of the Universe was in one place.

Then a huge explosion sent matter flying out in all directions. The particles of matter were spread out a long way from each other.

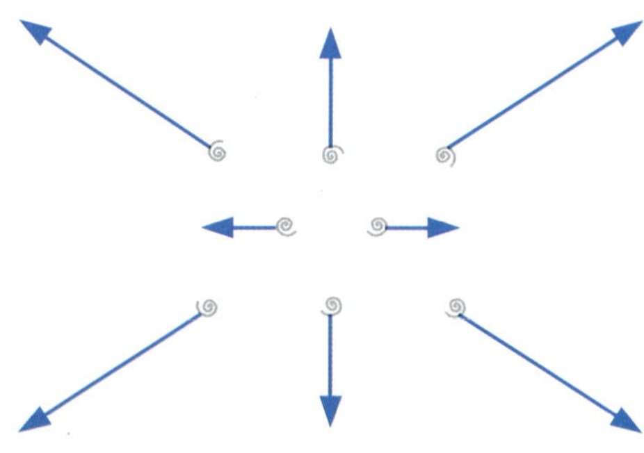

Gravity pulled the matter together in some places to make galaxies of stars. These galaxies are still moving away from each other.
The further away from each other galaxies are, the faster they are moving apart.

■ How do we know the Universe is expanding?

Astronomers examine the light from a star or a galaxy by splitting it up into a spectrum. They can tell from these spectra that distant galaxies are moving away from us.

The box explains how they do this.

> The spectra of stars and galaxies have patterns of dark lines. These are the 'fingerprints' of different elements.
>
> For example, in the yellow part of the spectrum of sunlight there is a pair of dark lines. These are due to sodium in the Sun's atmosphere.
>
> In the spectrum of the light from a distant galaxy, the lines of particular elements such as sodium are shifted towards the red end of the spectrum. This is called a **red-shift**. It tells us that the galaxy is moving away from us.
>
> A more distant galaxy has a bigger red-shift. This tells us that it is moving away faster.
>
> [Note: There are lots of dark lines spread right across the spectrum. Different lines correspond to different elements.]

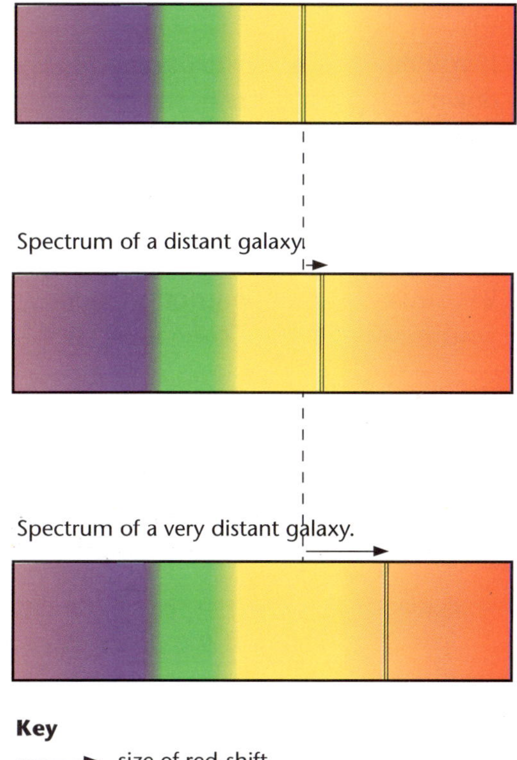

2 Copy and complete the sentences.

The light from distant galaxies shows a red-____________.
This tells us that the galaxies are moving ____________
____________ us.

More distant galaxies have ____________ red-shifts.
This tells us that they are moving away ____________.

Using your knowledge

F 10 The Universe may continue to expand for ever. Or it may eventually stop expanding and start to collapse again. This would end in a 'big crunch'. Explain each of these possibilities in terms of kinetic and potential energy.

F 11 The light from a few nearby galaxies shows a slight blue-shift. What does this tell us?

> **REMEMBER**
> When bodies move apart against the force of gravity, kinetic energy decreases and gravitational potential energy increases.

How to write good answers in GCSE Science examinations

■ Short answer questions

Here is an example of this type of question:

> The radiation from radioactive substances can be of three different types. Which of the three types of radiation is very short wavelength electromagnetic waves?

If you're not sure of the answer to this type of question, you might be tempted to write down _several_ answers in the hope that one of them is right. This is a bad idea. In most cases you'll automatically get _no_ marks.

If you wrote _gamma radiation or beta radiation_ as your answer to the above question you would score 0 marks even though gamma radiation is the correct answer.

You must make up your mind what you think is the most likely answer. If you change your mind later, you can cross out the old answer and write the new one.

■ Questions that require longer answers

It's easy to tell when you are expected to give a longer answer to a question. There will be lots of space for your answer and the question paper will indicate that you can score more than one or two marks for your answer. For example:

> Explain why the pressure of some air increases when you squeeze the air into a smaller space. (Assume that the temperature of the air does not change.)

This tells you the number of marks available. ⌐→ (4)
    └→ or (4 marks)

When answering these questions, don't just write down the first thing you think of and then leave it at that. You should include _all_ the relevant ideas that you can remember. But you should be sure that the ideas you write down really are relevant.

Don't write down things that you hope might just possibly be relevant. That's a sure way to lose marks because, if they're not relevant, it tells the person marking your answer that you don't really understand the question. This means that you will probably lose marks.

When answering the above question about the pressure of air, for example, you should not mention anything about the molecules in the air moving faster. This is because the question tells you that the temperature of the air stays the same. So the average speed of the molecules also stays the same.

You should also try to _organise_ your answer. This means putting all your ideas into a sensible order and then linking them together in a way that shows you really understand what's going on.

A few words like this in your answer:

First of all ... then ... because ... This means that ... So, on balance, ...

can help a lot.

Don't rush into writing down your longer answers.

Decide what the relevant ideas are and jot them down in pencil.

Then decide what order to write them down in and how you are going to link them together.

■ Science 'stories'

GCSE Science consists mainly of many separate ideas which, once you've understood them, you'll probably remember. But there are also some longer scientific 'stories' that you're also expected to remember, for example:

- explaining why things moving through a fluid (that is, a liquid or a gas) reach a terminal velocity;
- explaining why scientists believe that the Universe began with a 'big bang';
- explaining how we know about the internal structure of the Earth from P-waves and S-waves;
- describing the life histories of stars of different masses.

Very few candidates can <u>correctly</u> remember <u>all</u> the details of these stories that GCSE syllabuses require them to know.

Try setting out the stories in different ways, for example as a list of points in the correct order or in the form of a flow diagram. You will then find out which is the best way for <u>you</u> to remember them.

Finally, <u>practise</u> remembering the stories until you can remember them accurately.

■ Calculations

Physics questions on examination papers include a lot of calculations. So it's very important that you gain as many marks for them as you can. The formulas you are expected to know and/or be able to use are listed on pages 54–56 of this book.

Even if you get the wrong answer to a calculation, you can still get quite a lot of the marks. To gain these marks, you must have gone about the calculation in the right way. But the person marking your answer can only see that you've done this if you write down your working neatly and set it out tidily so it's quite clear what you have done.

You should:

- write down the formula you are using* (in words or symbols);
- put in the figures* for the quantities you know;
- calculate the quantity you have been asked to;
- write down the answer together with the units.

[* You may, at one of these stages, need to re-arrange the formula. Page 57 of this book tells you about how to use formulas.]

You should <u>not</u> alter incorrect figures. Instead, cross them out and write them again.

How to write good answers (continued)

> The current through a 12 V car headlamp bulb is 3 A.
> Calculate the resistance of the bulb.
>
> (3 marks)

You should set out your answer like this:

$watts = volts \times amps$ [or $power = p.d. \times current$]   ← Any one of these gets 1 mark
                 [or $P = V \times I$]   ← even if your answer is wrong.

$= 12 \times 3$

$= 36$ W [or $watts$]   ← The correct answer gets 2 marks.
                          ← The correct unit gets a mark,
                          even if your answer is wrong.

Always set out your classwork and homework calculations as in the example above so that you get into good habits. Then you'll still do calculations in the right way even under the pressure of examinations.

■ Two-stage calculations

You may not be able to calculate, in one step, what the question asks for. You may also need to calculate a value using one formula and then use this value in a second formula to calculate the answer to the question.

In these situations you should ask yourself the following questions:

- What information have I been given in the question?
- What values do I need to know to calculate the answer?
- Which of these values have I not been told?
- How can I calculate the unknown value from the information I am given?

> A 12 V car headlamp bulb has a resistance of 4 W. Calculate the power rating of the bulb.

$watts = volts \times amps \rightarrow$ But you don't know the amps, so you use $ohms = \dfrac{volts}{amps}$

$= 12 \times 3$ So amps $= \dfrac{volts}{ohms}$

$= 36$ W $= \dfrac{12}{4}$

                                      $= 3$ A
You can feed this into your original formula.

Get plenty of practice in doing two-stage calculations.
Then you'll get into the habit of asking yourself the right questions.

■ Interpreting graphs, etc.

When you are reading off values from a graph:

- check the scales on the axes so that you know what each small square on the grid represents;
- remember to quote units in your answer. [You can find these on the axis where you read off your answer. You can still quote the correct units even if you don't understand what they mean!]

Be as precise and accurate as you can:

- when describing trends or patterns
  [in the example, if you're asked what happens to the current when the voltage increases from 0.3 V to 0.5 V, 'increases steadily' is a better answer than 'increases'];
- when specifying key points
  [in the example:
  – current doesn't change until voltage reaches 0.3 V;
  – current starts to level off after 0.5 V;
  – current doesn't increase any more after voltage reaches 0.6 V;
  – maximum value of current is 0.9 A];
- when making comparisons
  [in the example, if you're asked how the current for 0.5 V compares with the current for 0.4 V, 'it's twice as big' or 'it's double the size' is a better answer than 'it's bigger'].

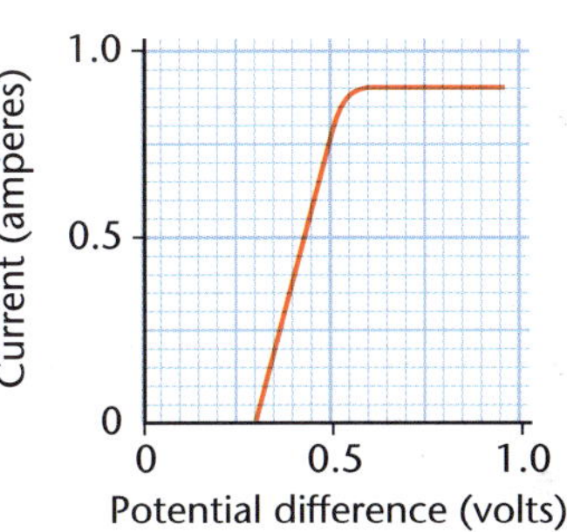

Get plenty of practice interpreting graphs, etc. as indicated above until you can do it perfectly with as little effort as possible.

■ Drawing graphs

- Choose sensible scales for the axes.
  [You should use <u>more than half</u> of the available squares along each axis.]
- Label the axes [e.g. Potential difference (volts)].
- Mark all the points neatly and accurately
  like this 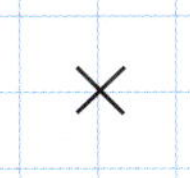 or like this 

- If the points are close to being a straight line or smooth curve, then draw the 'best fit' straight line or smooth curve. Use a pencil so you can rub your line out if you don't get it right first time. If there's an <u>obviously</u> wrong point, <u>ignore</u> it and indicate that you've done so.

Get plenty of practice doing these things so that you'll do the right thing even if you're nervous in an examination.

Physics formulas

■ Formulas that you must remember and be able to use

Energy

The cost of using an electrical appliance is given by:

cost = number of Units × cost per Unit

[Note: One Unit of electricity is a kilowatt-hour.]

The energy transferred by an electrical device can be calculated as follows:

$$\begin{array}{ccccc} \text{energy transferred} & = & \text{power} & \times & \text{time} \\ \text{(joules, J)} & & \text{(watts, W)} & & \text{(seconds, s)} \end{array}$$

Electricity

The rate of energy transfer by an electrical appliance (its power) is given by:

$$\begin{array}{ccccc} \text{power} & = & \text{potential difference} & \times & \text{current} \\ \text{(watts, W)} & & \text{(volts, V)} & & \text{(amperes, A)} \end{array}$$

The resistance of an electrical appliance or component is the number of volts of potential difference you have to apply across it for each ampere of current flowing through it. The formula relating resistance, potential difference and current is:

$$\text{resistance (ohms, } \Omega\text{)} = \frac{\text{potential difference (volts, V)}}{\text{current (amperes, A)}}$$

The voltages across the primary and secondary coils of a transformer are related as shown:

$$\frac{\text{voltage across secondary (volts, V)}}{\text{voltage across primary (volts, V)}} = \frac{\text{number of turns on secondary}}{\text{number of turns on primary}}$$

Waves and radiation

Wave-speed, wavelength and frequency are related as follows:

$$\begin{array}{ccccc} \text{wave-speed} & = & \text{frequency} & \times & \text{wavelength} \\ \text{(metres per second, m/s)} & & \text{(hertz, Hz)} & & \text{(metres, m)} \end{array}$$

For an object moving at a steady speed in a straight line:

$$\text{speed (metres per second, m/s)} = \frac{\text{distance travelled (metres, m)}}{\text{time taken (seconds, s)}}$$

For an object moving in a straight line with a steady acceleration:

$$\frac{\text{acceleration}}{\text{(metres per second squared, m/s}^2\text{)}} = \frac{\text{change in velocity (metres per second, m/s)}}{\text{time taken for change (seconds, s)}}$$

When a force moves an object, energy is transferred and work is done.

The work done (energy transferred), force and distance are related as shown:

work done = force applied × distance moved (in direction of force)
(joules, J)   (newtons, N)   (metres, m)

Pressure, force and area are related as shown:

$$\frac{\text{pressure}}{\text{(pascals, Pa)}} = \frac{\text{force (newtons, N)}}{\text{area (square metres, m}^2\text{)}}$$

(Notes: ■ One pascal is the same as one newton per square metre.
■ Other units of pressure, for example atmospheres, are often more convenient.
You should use whatever units you are given in a particular question.)

An unbalanced force acting on an object causes it to accelerate.
Force, mass and acceleration are related as shown:

force   =   mass   ×   acceleration
(newtons, N)   (kilograms, kg)   (metres per second squared, m/s^2)

The pressure and volume of a fixed mass of gas at a constant temperature
are related as shown:

initial pressure × initial volume = final pressure × final volume

(Note: You should use whatever units you are given in a particular question.)

There is more about physics formulas on the next two pages.

Physics formulas (continued)

■ **Formulas that you must be able to use but do <u>not</u> need to remember**

Energy

Energy transferred from the mains supply is measured in kilowatt-hours or Units:

$$\text{energy transferred} = \text{power} \times \text{time}$$
(kilowatt-hours, kW h)   (kilowatts, kW)   (hours, h)

The power of an energy-transferring device is calculated as follows:

$$\text{power (watts, W)} = \frac{\text{energy transferred (joules, J)}}{\text{time taken (seconds, s)}}$$

When an object is lifted (or falls), potential energy is transferred to (or from) it:

$$\text{change in gravitational potential energy} = \text{weight} \times \text{change in vertical height}$$
(joules, J)   (newtons, N)   (metres, m)

You can calculate the efficiency of an energy-transferring device as follows:

$$\text{efficiency} = \frac{\text{useful energy transferred by device}}{\text{total energy transferred by device}}$$

(Note: An efficiency is just a number; it has no units. You can state an efficiency as an ordinary fraction, a decimal fraction or a percentage e.g. $\frac{3}{4}$, 0.75 or 75%.)

Electricity

The electrical charge that flows depends on current and time as shown:

$$\text{charge} = \text{current} \times \text{time}$$
(coulombs, C)   (amperes, A)   (seconds, s)

Voltage is a measure of the energy transferred when 1 coulomb of charge flows:

$$\text{energy transferred} = \text{potential difference} \times \text{charge}$$
(joules, J)   (volts, V)   (coulombs, C)

Forces

The weight of an object is the force of gravity that acts on it:

$$\text{weight} = \text{mass} \times \text{gravitational field strength}$$
(newtons, N)   (kilograms, kg)   (newtons per kilogram, N/kg)

The kinetic energy of a moving object is given by:

$$\text{kinetic energy} = \frac{1}{2} \times \text{mass} \times (\text{speed})^2$$
(joules, J)   (kilograms, kg)   [(metres per second)^2, (m/s)^2]

■ Using physics formulas

- When you are asked to do a physics calculation, you should begin by writing down the formula you need to use.

For example, to calculate the amount of electrical charge that flows you write:

$$charge = current \times time \qquad \text{or} \qquad coulombs = amperes \times seconds$$

[This is really the correct way.]   [This is allowed in GCSE; it tells you the units.]

- Sometimes you can use a formula just as it is.

<u>Example</u>   What is the power of a 12 V, 3 A car headlamp bulb?

$$power = potential\ difference \times current$$
$$= 12 \times 3$$
$$= \underline{36\ watts}\ (or\ W)$$

- Sometimes you need to change one (or more) of the units.

<u>Example</u>   How much energy is transferred by a 100 W light bulb in 5 minutes?

$$energy\ transferred = power \times time \qquad\qquad time = 5\ minutes$$
$$(in\ seconds) \qquad\qquad\qquad = 5 \times 60\ seconds$$
$$= 100 \times 300 \qquad\qquad\qquad \leftarrow\ = 300\ seconds$$
$$= \underline{30\,000\ joules}\ (or\ J)$$

- Sometimes you need to do some re-arranging.

<u>Example</u>   What force gives a pressure of 10 Pa on an area of 3 m^2?

You can re-arrange the formula first: or   You can put in the numbers you know:

$$pressure = \frac{force}{area} \qquad\qquad\qquad 10 = \frac{force}{3}$$

So

$$pressure \times area = force$$
$$10 \times 3 = force$$
$$force = \underline{30\ newtons}\ (or\ N)$$

and **then** re-arrange the formula
$$force = 3 \times 10$$
$$= \underline{30\ newtons}\ (or\ N)$$

> Here is another re-arrangement you might sometimes need:
>
> $$pressure = \frac{force}{area}$$
>
> So area $= \dfrac{force}{pressure}$

Coverage of GCSE Science syllabuses

Though the GCSE Science syllabuses from Edexcel, MEG, NEAB and SEG are very similar, there a few small differences. Together, *Science Foundations Physics* and *Extension Physics* cover <u>all</u> the knowledge and understanding that is required by Higher-tier candidates entered for NEAB syllabuses (Coordinated or Modular). They also cover <u>most</u> of the further knowledge specified for Higher-tier candidates by other examining groups. Items of Higher-tier content from these other syllabuses that are <u>not</u> included in *Science Foundations Extension* are listed below.

■ Energy

None

■ Electricity

Use of Fleming's left-hand rule to relate/predict the direction of the force/movement of a current-carrying wire in a magnetic field. (Edexcel; SEG)

Understand that the principle of electromagnetic induction can be used in the detection of metal pipes and wires in walls. (Edexcel)

Appreciate how LDRs and thermistors can be used with electrical circuits to monitor light levels and temperature in a building. (MEG)

Understand how charging by induction occurs in terms of the movement of electrons. (MEG)

■ Waves and radiation

Use the quantitative relationship:

$$\text{frequency} = \frac{1}{\text{time period}}$$
$$f = \frac{1}{T}$$

(Edexcel)

Understand that the extent of diffraction depends on wavelength <u>and the size of the obstacle and/or aperture including the effect of doorways on sound waves and narrow slits on light waves.</u> (Edexcel)

Know that water waves can spread out at a <u>narrow</u> gap and that this is known as diffraction.
Understand how the amount of spreading depends on the size of the gap compared with the wavelength of the wave.
Know that light can be diffracted <u>but needs a very small gap as the wavelength of light is very small</u>. (MEG)

Understand that optical fibres allow the rapid transmission of data using digital signals.
Understand how information in narrow beams can be transmitted using microwaves. (MEG)

■ Forces

Recall evidence for the 'big bang' theory <u>including the background microwave radiation</u>. Consider other theories of the origin of the Universe such as the 'steady state' theory. (Edexcel)

Asteroids as a group of rock debris occupying a belt between the orbits of Mars and Jupiter. Comets as bodies orbiting the Sun <u>in different planes from that of the planets</u>. (SEG)

Use the equation:

$$\text{energy transferred} = \text{force} \times \text{distance} = \tfrac{1}{2}mv^2$$

<u>to discuss stopping distances.</u>

Apply the relationship:

$$\text{force} = \text{mass} \times \text{acceleration}$$

<u>to the action of seat belts and crumple zones.</u> (MEG)

Suggestions for practical work

The needs of Higher-tier students for appropriate practical work can often be met by providing them with more demanding briefs for the practical work suggested in *Science Foundations Physics Supplementary Materials*.

Further practical activities are also often suggested by the *Science Foundations Extension* topics themselves.

Teachers may, however, find *Physics Activities for GCSE* (Nuffield Foundation, 1996) a useful additional resource. The following table matches each *Extension* topic against the activities in this resource.

Science Foundations Extension topic		*Physics Activities for GCSE* reference
Energy	H1	P61, P62, P63
	H2	P67
	H3, H4	P97(b)
Electricity	H1	P22, P23, P24
	H2	P93, P94
	H3	P95
	H4	P96
	H5	P55
	H6	P26, P28
	H7	P48
Waves and radiation	H1	P38, P40
	H2	P78, P79
	H3	P71
	H4	P43 [extends beyond diffraction at a single edge]
	H5	P74, P77
	H6	P76, P80
Forces	H1	P14, P15
	H2	P89
	H3	P83, P84
	H4	P86(a)
	H5	P3, P4, P6, P87
	H6	P88
	H7	P86(b)

W 12

(a)

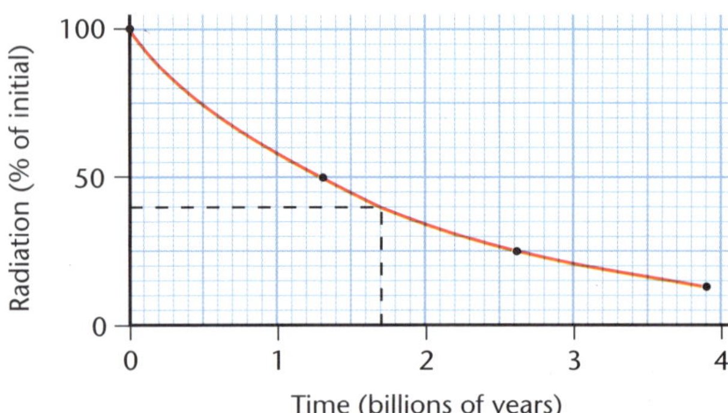

(b) 2 atoms of K-40 to 3 atoms of Ar-40 means that only 40% of the original K-40 remains and that 60% has decayed. From the graph, this would have taken 1.7 billion years.

F 2

The particles (or molecules) are spread out further so they hit the sides of whatever contains them less often. This means that there is less pressure on the container.

EI 10

(a) charge = current × time [in seconds]
$$= 0.5 \times (2 \times 60 \times 60)$$
$$= 0.5 \times 7200$$
$$= 3600 \text{ coulombs (or C)}$$

(b) The mass of silver produced is in proportion to current × time. [Current is scaled by $\frac{0.2}{0.5}$ and time is scaled by $\frac{1.5}{2}$.]

So mass of silver $= 8 \times \frac{0.2}{0.5} \times \frac{1.5}{2}$
$$= 2.4 \text{ grams (or g)}$$

En 4

Global warming is caused mainly by the greenhouse effect rather than by the energy released from fuels. Nuclear fuels do not produce waste gases and so do not contribute to global warming. Energy in the wind in any case ends up as low-grade thermal energy, so intercepting it makes no difference at all to global warming. [Note: Building nuclear power stations and wind generators does contribute to global warming.]

F 9

The molecules of gases such as oxygen, nitrogen, carbon dioxide and water vapour must all have average speeds greater than 2.4 km/s (at the temperature on the Moon) so that they all escape from the Moon.

W 1

(a) Radio 4 long wave has waves that are $\frac{1500}{3} = 500$ times longer than Radio 4 FM. So the frequency of Radio 4 FM is 500 times greater than Radio 4 long wave.

(b) wave-speed = frequency × wavelength
$$= 100 \text{ million} \times 3$$
$$= 300 \text{ million metres per second}$$
$$[300\,000\,000 \text{ m/s or } 3 \times 10^8 \text{ m/s}]$$

EI 7

(a) $\frac{400\,000}{230} = 1739$ (to the nearest whole number)

So the current would be 1739 times greater (for the same power).

(b) Power loss increases as square of current. So it is:
$$(1739)^2 = 3\,024\,121 \text{ times greater}$$
[more than 3 million times greater].

W 8

The VHF radio signals have a longer wavelength than the UHF TV signals and are more strongly diffracted by the top of the hill.

F 5

(a) The bigger the mass of a star, the shorter its lifetime as a stable star like the Sun (what we call a **main sequence** star). Halving the mass of a star more than doubles its lifetime.

(b) The bigger the mass of a star, the larger the gravitational forces pulling it together and the higher the temperature needed to keep it stable. The higher temperature makes the nuclear fusion reactions occur faster. This increase in the rate of the fusion reactions is bigger than the increase in the mass that causes it, so all the hydrogen is used up sooner.

En 7

(a) [Note: g.p.e. stands for 'gravitational potential energy'.]

$$\text{Change in g.p.e.} = \text{weight} \times \text{change in height}$$
$$= 10\,\text{N} \times 2\,\text{m}$$
[1 kg has a weight of 10 N on Earth]
$$= 20\,\text{J}$$

(b)
$$\text{Change in g.p.e.} = \text{weight} \times \text{change in height}$$
$$= 500\,\text{N} \times 1500\,\text{m}$$
[50 kg]
$$= 750\,000\,\text{J}$$

El 5

(a) Very strong electromagnets are an advantage because the voltage induced in the coils will be greater.

(b) The problem with using the generator to supply the current for its own electromagnets is getting the generator to produce a current in the first place. [In practice, this isn't a problem because the iron core of the electromagnet always stays slightly magnetised. This is enough for the generator to <u>start</u> producing electricity which then activates the electromagnet so that more electricity is produced.]

W 3

(a) and **(c)**

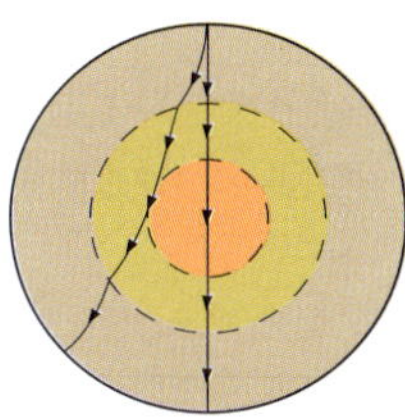

(b) The P-wave shown does not pass through the solid inner core.

F 7

The gradient of the velocity–time graph becomes less and less steep until it has zero gradient, i.e. it is horizontal. This means that the acceleration is becoming smaller and smaller until the object is travelling at a constant speed. [The graph could represent a falling object (or a vehicle) reaching its terminal velocity.]

En 2

(a)
Cost of installing gas fire = £400
Annual saving in fuel = £200
Pay-back time = 2 years

(b) This pay-back time is longer than for draught excluders or loft insulation but is shorter than for cavity wall insulation or double glazing. The gas fire should last for a lot longer than 2 years, so it is a good investment.

(c) Gas fires are better than coal fires because:

- they do not produce smoke;
- the waste gases do not cause acid rain [this is because they contain hardly any sulphur dioxide; burning <u>any</u> fuel, however, produces some nitrogen oxides, which also contribute to acid rain];
- the waste gases contain less of the greenhouse gas carbon dioxide (in relation to the energy released);
- they are more efficient, i.e. more of the thermal energy is transferred to the room rather than escaping up the chimney.

El 11

(a)
$$\text{energy transferred} = \text{p.d.} \times \text{charge}$$
$$= 12 \times 500$$
$$= 6000\,\text{J (or joules)}$$

(b)
$$\text{energy transferred} = \text{p.d.} \times \text{charge}$$
So
$$\text{p.d.} = \frac{\text{energy transferred}}{\text{charge}}$$
$$= \frac{9600}{240}$$
$$= 40\,\text{V (or volts)}$$

F 1

(a) The air is squeezed into $\frac{100}{35} = 2.86$ times less space. So you would expect the pressure to be 2.86 times greater than atmospheric pressure (i.e. 2.86 atmospheres).

(b) The measured pressure is higher than expected so the air must have become hotter as it was compressed. As it cools down, the pressure falls to what was expected.

Answers (2)

En 8

(a) efficiency $= \dfrac{\text{useful energy transferred}}{\text{total energy transferred}}$

$= \dfrac{750\,000}{1\,000\,000}$ ← [from 1st example on page 13]

$= 0.75$ or 75%

(b) efficiency $= \dfrac{\text{useful energy transferred}}{\text{total energy transferred}}$

$= \dfrac{750\,000}{1\,250\,000}$ ← [from 2nd example on page 13]

$= \dfrac{3}{5}$ (or 0.6 or 60%)

W 6

speed $= \dfrac{\text{distance}}{\text{time}}$

So

distance = speed × time

$= 1500 \times 0.0001$

⌐ [0.1 milliseconds = 0.0001 s]

$= 0.15$ metres (or 15 cm)

But this is the distance through the metal <u>and back again</u>. So the thickness of the metal is $\frac{15}{2}$ = 7.5 cm.

F 10

As galaxies move apart against the force of the gravitational attraction between them, kinetic energy is transferred as gravitational potential energy. The force of gravity between galaxies becomes smaller as they get further apart. So if the galaxies have enough kinetic energy, this will never all be transferred as gravitational potential energy and the galaxies will continue to move apart.

However, if <u>all</u> the kinetic energy of the galaxies is transferred as gravitational potential energy, the Universe will stop expanding and start to contract. As it does so, gravitational potential energy will be transferred as the increasing kinetic energy of the galaxies and there will eventually be a 'big crunch'.

En 1

(a) Objects above −273 °C emit energy. The rate at which they emit energy increases as the temperature increases. It increases <u>more</u> than in proportion to the increase in temperature. [Note: −273 °C is as cold as you can get. It is the **absolute zero** of temperature, or zero on the **kelvin** scale. So 0 °C is 273 kelvin (273 K).]

(b) The wavelength at which most energy is emitted becomes shorter as the temperature increases. [You can, for example, feel the infrared radiation emitted from a body that isn't hot enough to emit light.]

El 2

(a) [from the graph] 250 Ω (or ohms)

(b) [from the graph] 4 kΩ (or 4000 ohms)

W 9

Light waves have a much shorter wavelength than radio waves.

F 8

Kinetic energy is proportional to the square of the speed. So a car that moves $\frac{70}{50}$ times as fast has $\left[\frac{70}{50}\right]^2 = 1.96$ times as much kinetic energy. This means that there is almost twice as much chance of being killed.

En 3

(a) To keep it fresh, the air inside the house needs to be changed every couple of hours. Using a heat exchanger to do this means that the incoming air is heated up by the outgoing air so that less thermal energy is lost to the surroundings.

(b) Cost $= £2000$

Annual saving $= £\,400$

Pay-back time $= 2000 \div 400$

$= 5$ years

(c) The heat exchanger system is about as cost effective as cavity wall insulation and more cost effective than double glazing. Assuming a lifetime of say 20 years, the heat exchanger system would save £6000.

EI 1

$$\text{resistance} = \frac{\text{potential difference}}{\text{current}}$$

$$= \frac{10}{2}$$

$$= 5\,\Omega \text{ (or ohms)}$$

W 5

The pulses labelled X and Y are:

- nearer to pulse W than B and C are to A;
- nearer to each other than pulses B and C.

This means that this flaw in the casting is nearer to the front surface and is narrower than the other flaw.

F 4

(a)

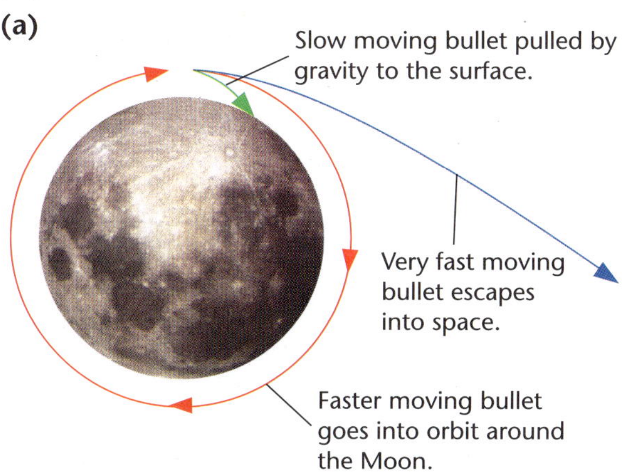

(b) You can't make an object go into orbit at the height of the highest mountain on Earth because air resistance would constantly slow it down.

EI 8

(a) If current increases 1000 times, the power loss in the fuse increases $(1000)^2$ times = 1 million times.

(b) The power loss in the fuse is due to energy being transferred as thermal energy. Normally, this makes the fuse very slightly warm. Transferring a million times more thermal energy than normal in the fuse makes it very hot and melts it.

W 13

(a) The scientists could put a small amount of a soluble radionuclide on to the waste dump then test the water in the river over a period of time to see if it contains any of the radionuclide.

(b) The radionuclide should have a half-life that is long enough for the tests to take place but short enough for it not to remain very radioactive for very long after the experiment is completed. Depending on the site, a half-life of a few days to a few months would be suitable.

Only a small amount of radionuclide would be used, so the greatest danger to living organisms would be if it got <u>inside</u> them. This means that the radionuclide should <u>not</u> be an alpha emitter. Beta or gamma radiation would also be easier to detect.

EI 3

$$\text{resistance} = \frac{\text{potential difference}}{\text{current}}$$

So

$$\text{current} = \frac{\text{potential difference}}{\text{resistance}}$$

$$= \frac{230}{920}$$

$$= 0.25\,\text{A (or amperes)}$$

F 11

The blue-shifts tell us that the Earth and these nearby galaxies are moving <u>towards</u> each other.

En 5

Pollution doesn't have to involve <u>chemicals</u>; it refers to anything in the environment that people would prefer not to be there. Many people do not like to see hill-tops covered with large wind generators. They think these wind-farms are an eye-sore, so that they cause <u>visual</u> pollution.

Answers (3)

En 6

If there is an accident at a nuclear power station (such as several small accidents at Sellafield in the UK and the large accident at Chernobyl in the Ukraine in 1986), the radioactive materials that are released can harm living things, including humans. There is also the problem of making sure that wastes which can remain dangerously radioactive for thousands of years are stored safely and don't escape into the environment.

W 4

P-waves do not produce a shadow zone. This is because they can travel through the core of the Earth.

F 3

(a) Area of slave piston is $\frac{75}{1.5} = 50$ times greater than area of master piston. There is the same pressure throughout the liquid, so on the slave piston there is 50 times as much force as on the master piston. This means that the force on the master piston is $\frac{2500}{50} = 50$ newtons (or N).

(b) pressure $= \dfrac{\text{force}}{\text{area}}$

$\quad = \dfrac{2500}{75}$ or $\dfrac{50}{1.5}$

$\quad = 33.33\,\text{Pa}$ (or pascal or N/m^2)

El 4

The diode has a very high (infinite!) resistance when a potential difference is applied across it in one direction. This is also the case for potential differences up to 0.3 V applied across it in the opposite direction. For potential differences greater than 0.3 V in this opposite direction, the diode has a low(er) resistance.

[You can't calculate the actual resistance for any particular voltage because there isn't a scale on the current axis of the graph.]

W 2

Blue light and red light travel at the same speed [in a vacuum and very nearly at the same speed in air]. So if blue light has double the frequency of red light it has only half the wavelength of red light.

El 6

(a) $\dfrac{\text{primary voltage}}{\text{secondary voltage}} = \dfrac{\text{primary turns}}{\text{secondary turns}}$

$\dfrac{25\,000}{400\,000} = \dfrac{\text{primary turns}}{\text{secondary turns}}$

$\dfrac{1}{16} = \dfrac{\text{primary turns}}{\text{secondary turns}}$

So the secondary coil has 16 times as many turns as the primary coil.

(b) $\dfrac{\text{primary voltage}}{\text{secondary voltage}} = \dfrac{\text{primary turns}}{\text{secondary turns}}$

$\dfrac{230}{\text{secondary voltage}} = \dfrac{1}{16}$

$\qquad$ secondary p.d. $= 16 \times 230$

$\qquad\qquad\qquad = 3680\,\text{V}$ (or volts)

[Alternatively you could simply say that the transformer now steps up the voltage 16 times: $230 \times 16 = 3680\,\text{V}$]

W 7

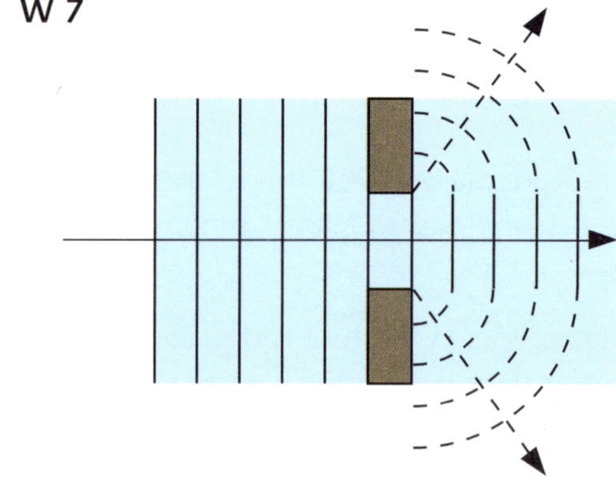

W 11

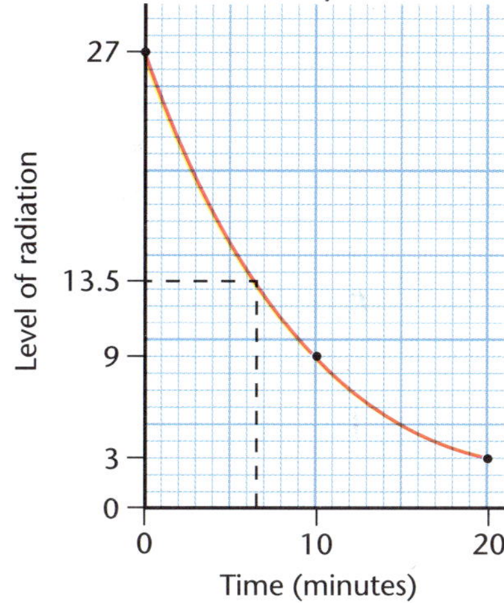

The graph shows that the half-life is about 6.5 minutes.

EI 9

The volume of hydrogen produced is in proportion to current × time. So the volume produced is
$20 \times 2 \times 5 = 200\,\text{cm}^3$
[current scaled up by 2, time scaled up by 5]

F 6

The gradient on the distance–time graph becomes steeper and steeper. This means that the object is moving faster and faster, i.e. it is accelerating.

W 10

(a) After 100 years the radionuclide has only $\frac{1}{2}$ of its original activity. In another 100 years it has only $\frac{1}{2} \times \frac{1}{2} = \frac{1}{4}$ of its original activity. So the answer is 200 years.

(b) When a sample of radionuclide is 10% as radioactive as it once was, only 10% of the original unstable atoms are left. This means that 90% of them have already decayed.

EI 12

The explanation is on the final part of page 23, starting from 'In other words, …'

Additional questions: Energy

You may need to use any of the formulas under 'Energy' on pages 54 and 56.

1 The diagrams show what happens to each 100 J of energy from a fuel when it is used in a vehicle with a diesel engine and when it is used to supply the electricity for an electrically powered vehicle.

(a) Make a copy of each diagram and add the missing figures.

(b) What is the efficiency:

   (i) of the diesel engine;

   (ii) of the motor in the electric vehicle;

   (iii) of the electric vehicle overall?

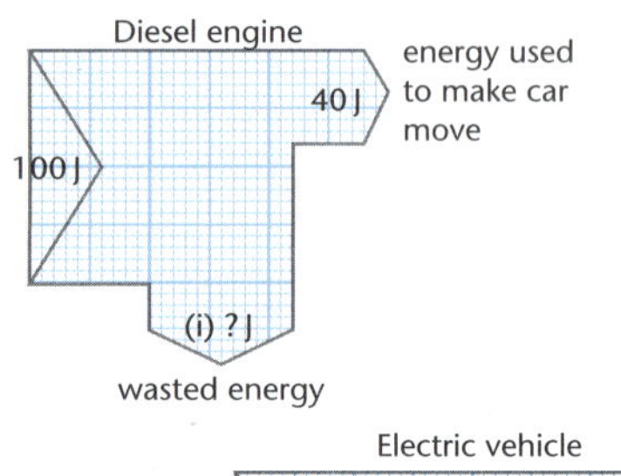

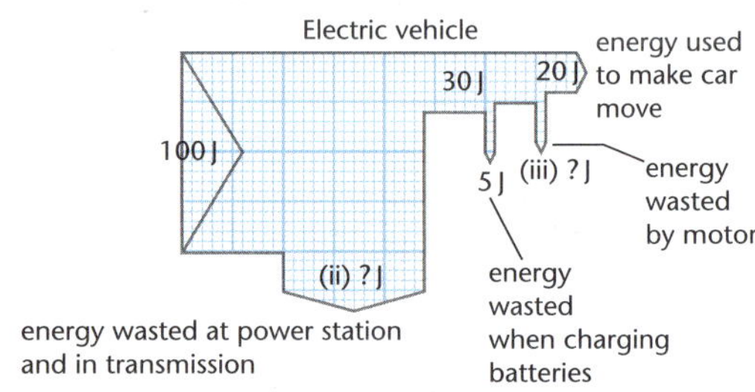

2 The photographs show two light bulbs that give out exactly the same amount of light.

	New	Old
Input power	20 W	100 W
Cost (for 5000 hours) ■ to buy ■ to use	£7.00 £8.00	5 × 30p £40.00

Compare the two bulbs in terms of:

(a) their efficiency;

(b) their cost-effectiveness.

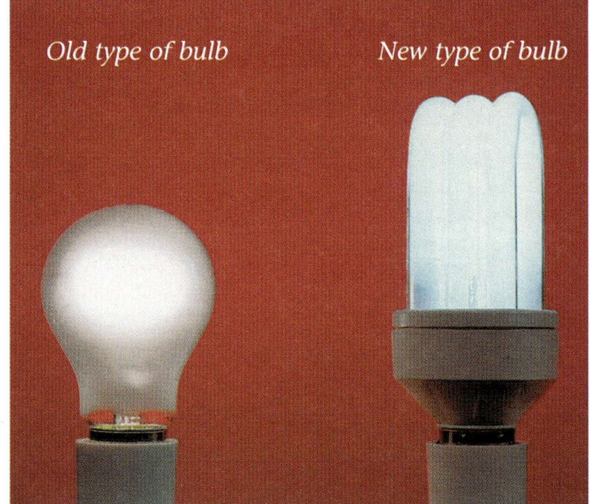

3 Glider pilots often gain height by using rising currents of air, called thermals, over large areas of tarmac or recently ploughed fields.

(a) Explain:

   (i) why the air above such areas becomes warmer than in other places on sunny days;

   (ii) why the warm air rises.

(b) Calculate the increase in gravitational potential energy of the glider and its pilot when they climb 100 metres.

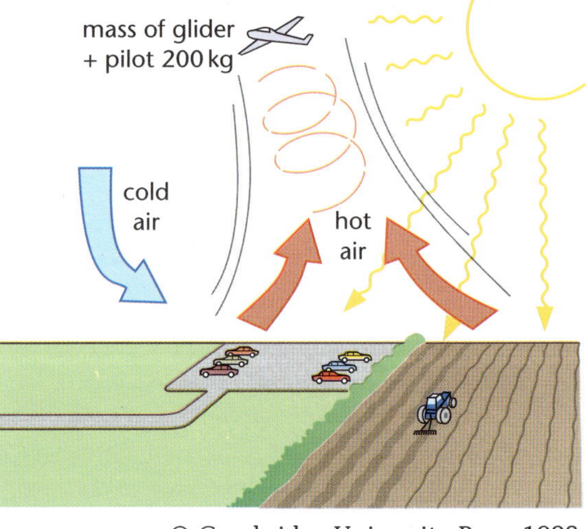

4 Burning fuels produce waste gases that escape into the atmosphere.

Study the information on the bar-charts and pie-chart, then use it to answer the following questions.
[Give reasons for each of your answers.]

(a) Which method of heating a room will result in the most acid rain being produced?

(b) Which method of heating a room will produce the smallest increase in the greenhouse effect?

(c) In what <u>two</u> ways does using natural gas as a fuel cause an increase in the greenhouse effect?
[You may now want to revise your answer to (b)]

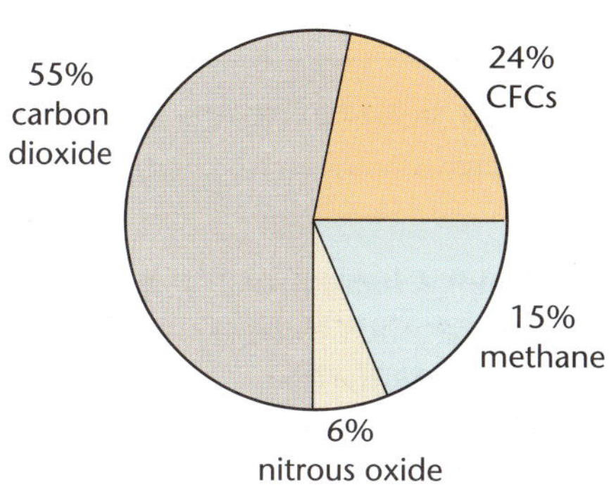

Emissions caused by heating a living room

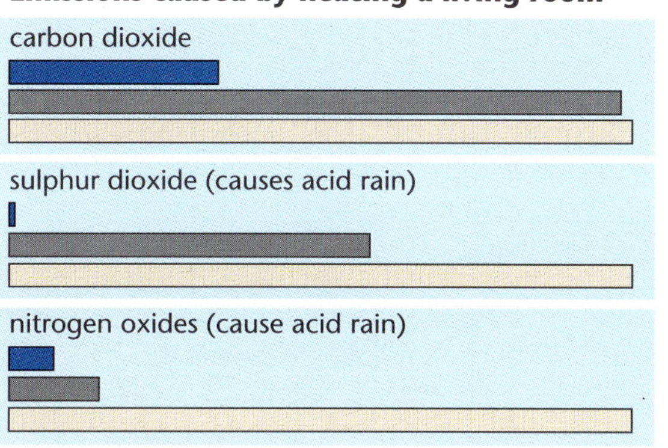

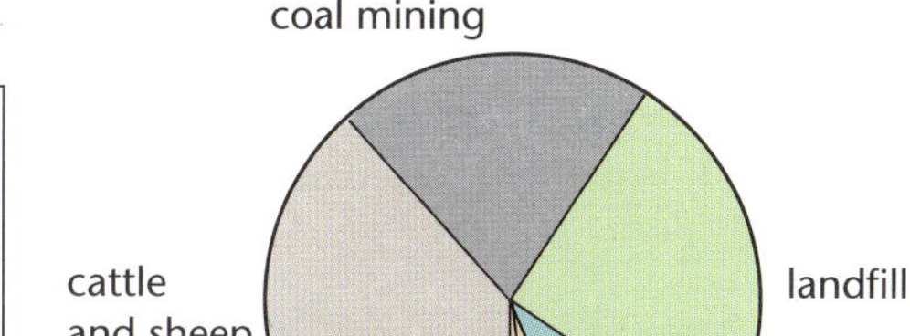

5 A new type of solar energy collector has recently been designed. The new type of collectors are much more efficient than the usual 'flat plate' collectors.

Study the information about the new type of solar energy collector, then answer the questions below.

(a) What are the advantages of having the collector plate and heat pipe inside an evacuated tube?

(b) How is the collector plate made very efficient?

(c) How is thermal energy transferred from the collector plate to water so that it can be used inside a house?

(d) Why is the heat pipe described as a diode?

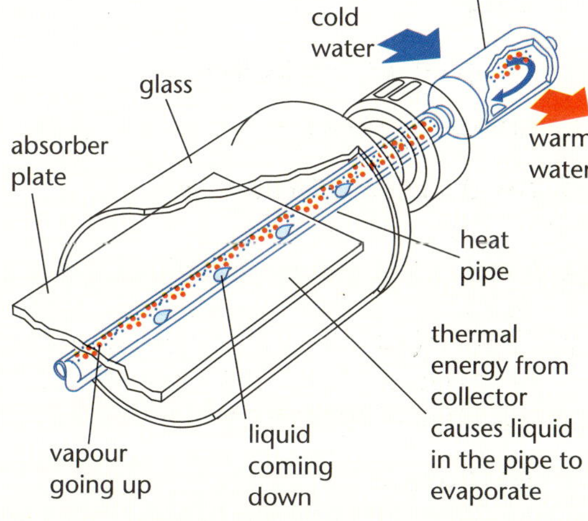

The collector plates are sealed inside evacuated tubes to eliminate thermal energy losses due to conduction and convection. They are also protected against damp and corrosive gases in the air.
The collector plates have a special coating which absorbs the maximum amount of solar radiation but minimises the energy that is re-radiated at much longer wavelengths.
Transfer of thermal energy from the collector plates to water is via heat pipes which contain a liquid.
The diagram shows how these heat pipes work. The heat pipes also act as diodes; in other words, thermal energy transfer is always from the collector plates to the water and never in the reverse direction.

Additional questions: Electricity

You may need to use any of the formulas under 'Electricity' on pages 54 and 56.

1 When a filament lamp is switched on, it takes a few milliseconds to heat up to its operating temperature of about 2500 °C.

The graph shows what happens to the current through the filament of a 230 V mains lamp as it heats up to its operating temperature.

(a) How does the initial current through the filament compare with the current at its operating temperature?

(b) Calculate the resistance of the filament:

    (i) at 0 °C;

    (ii) at 1000 °C;

    (iii) at 2500 °C.

(c) Use the information from the second graph to explain why the filament reaches a maximum temperature.

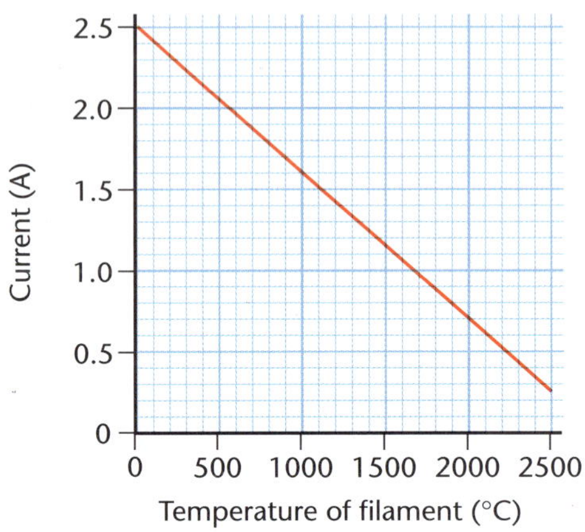

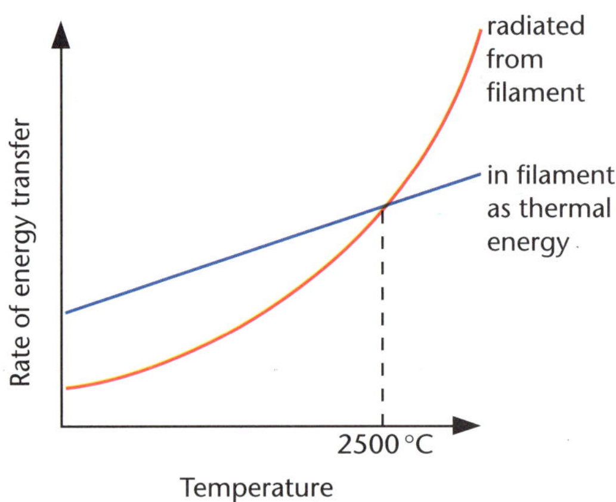

2 The diagram shows a transformer whose primary coil is connected to a d.c. supply.

Nothing happens in the secondary circuit until the current in the primary circuit is switched off. Then a spark is produced between the two points.

(a) Why does nothing happen in the secondary circuit when a steady current flows through the primary circuit?

(b) Explain, as fully as you can, why you get a spark across the points connected to the secondary coil when the current through the primary is switched off.

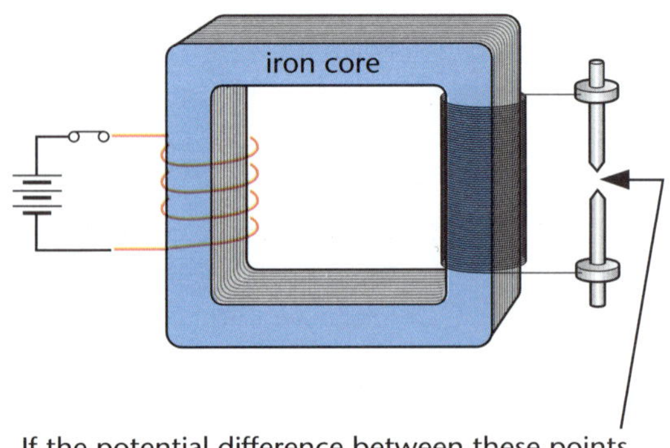

If the potential difference between these points is big enough, a spark is produced.

3 The diagram shows the electrolysis of molten aluminium oxide. This is how aluminium metal is produced.

(a) An electric current is a flow of electrically charged particles. What type of electrically charged particles are moving:

    (i) in the wires that carry the current to and from the electrodes;

    (ii) in the molten aluminium oxide?

(b) In which directions do the charged particles in the molten aluminium oxide move?

(c) A current of 100 A for 15 minutes produces 9 g of aluminium metal.

    (i) How many coulombs of charge must flow for each gram of aluminium produced?

    (ii) How much aluminium is produced when a current of 150 A flows for 10 minutes?

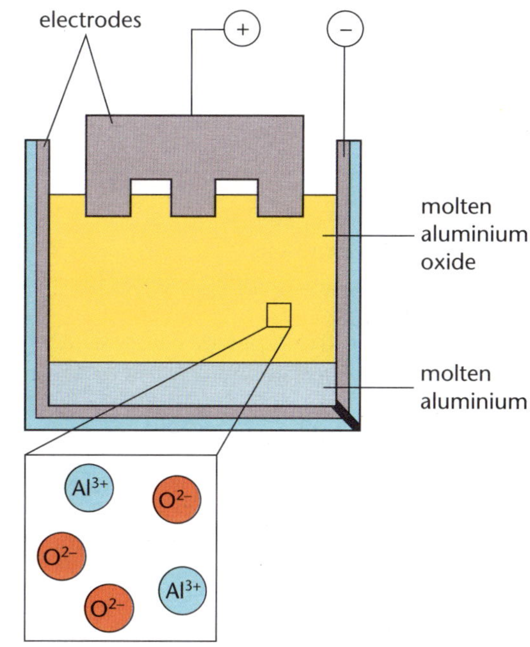

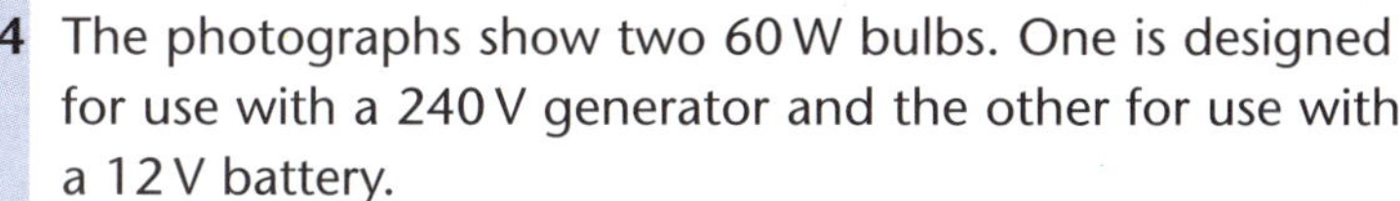

4 The photographs show two 60 W bulbs. One is designed for use with a 240 V generator and the other for use with a 12 V battery.

(a) Calculate the current that flows through each bulb.

(b) The cable to each bulb has a resistance of 0.1 Ω. Calculate the power loss in the cable in each case.

$$\text{Power loss} = (\text{current})^2 \times \text{resistance}$$
$$\quad\ \ (\text{W}) \qquad\quad (\text{A}) \qquad\qquad (\Omega)$$

5 A woman walks across a dry carpet, sliding the soles of her shoes. Just before she touches a water tap, a 25 000 V spark jumps across from her hand. This doesn't harm her in any way.

If the same woman touched the live wire of the 230 V mains, she could be killed.

Calculate in each case:

(a) the current that flows;

(b) the energy that is transferred.

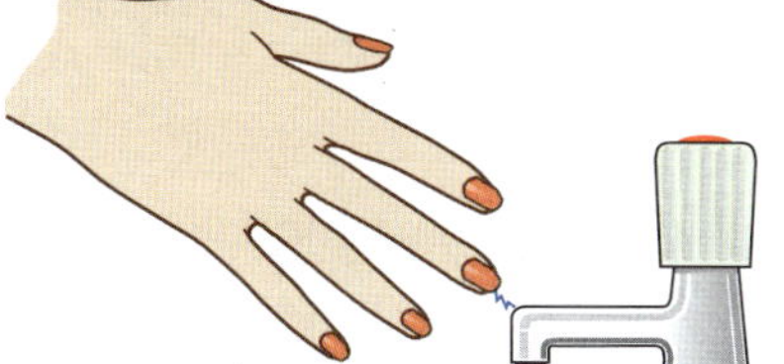

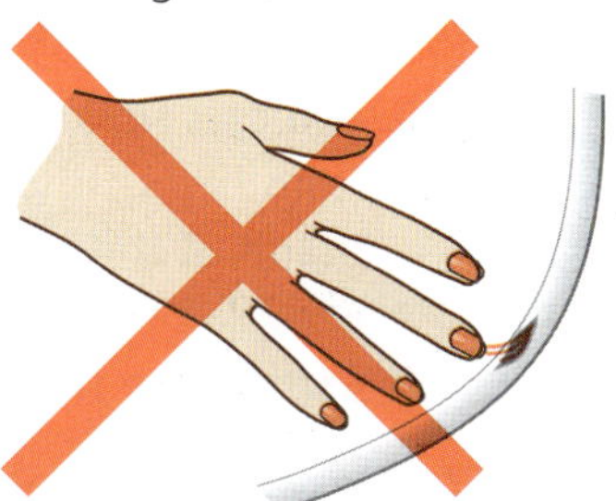

To make the spark, a billionth of a coulomb of charge flows in a hundredth of a second.

In this electric shock, a hundredth of a coulomb of charge flows in a tenth of a second.

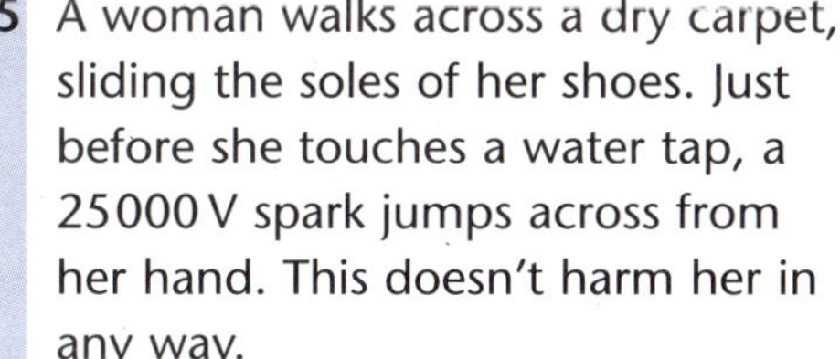

Additional questions: Waves and radiation

You may need to use the formula under 'Waves and radiation' on page 54.

1 The diagrams X and Y show two different ways of sending waves along a spring.

  (a) Name the types of waves in X and Y. Give reasons for your answers.

  (b) What is the wavelength of each type of wave?

  (c) Both waves have a frequency of 2 Hz. What is the speed of each type of wave?

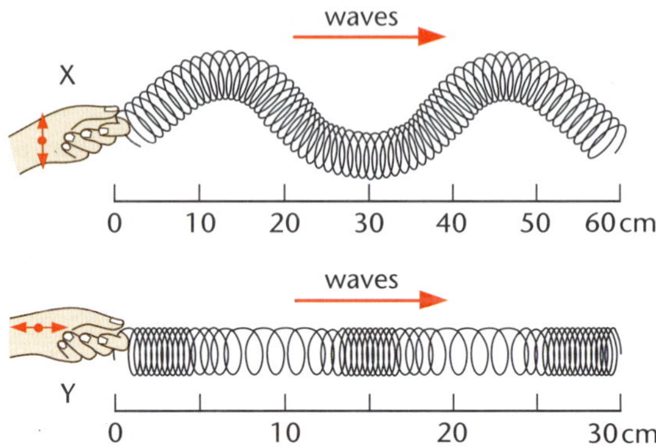

2 VHF (very high frequency) radio signals have a frequency of around 100 MHz.

UHF (ultra high frequency) TV signals have a frequency of around 1000 MHz.

  (a) How do the wavelengths of the VHF and UHF signals compare?

  (b) Some houses are in the radio shadow of a block of flats. Which of the signals are the houses most likely to be able to receive? Give a reason for your answer.

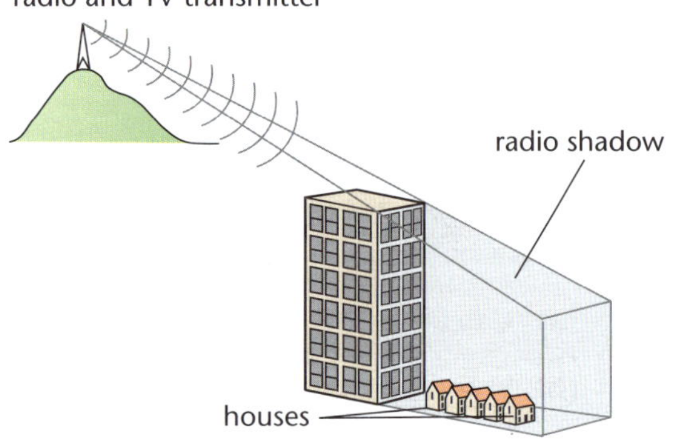

3 The diagram shows a seismograph trace of the P-waves and S-waves that have travelled through the Earth's crust from an earthquake.

  (a) All that you know from the trace is the time gap in the arrival of the two types of waves. What is this time gap?

  (b) You can work out how far away the earthquake is from the seismograph station using a distance–time graph of the two types of waves. The way to do this is explained on the graph. How far away is the earthquake shown on the seismograph trace?

  (c) What is the speed through the Earth's crust of each type of wave?

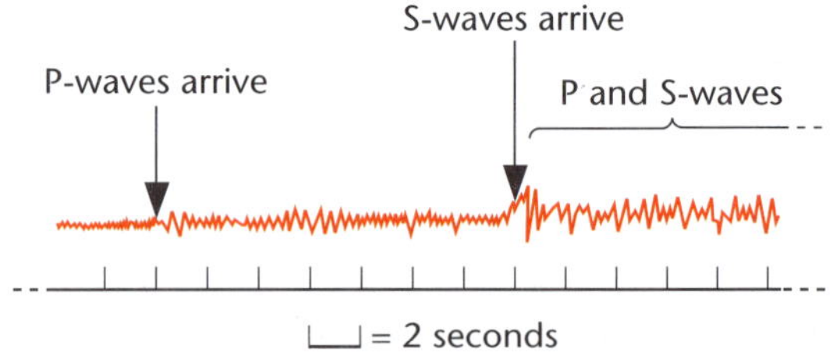

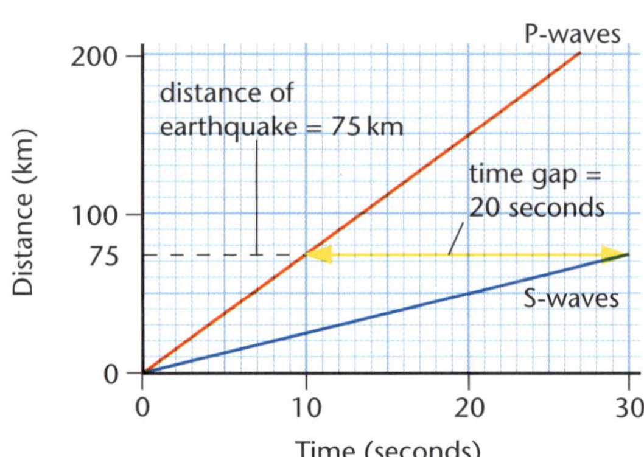

© Cambridge University Press 1998

4 Pulses of ultrasound can be used to measure the thickness of the fat on a pig's back.

(a) What do the pulses A, B and C on the ultrasound trace represent?

(b) How thick a layer of fat does the ultrasound trace show?

(c) Suggest why there are other small pulses, D and E, on the ultrasound trace?

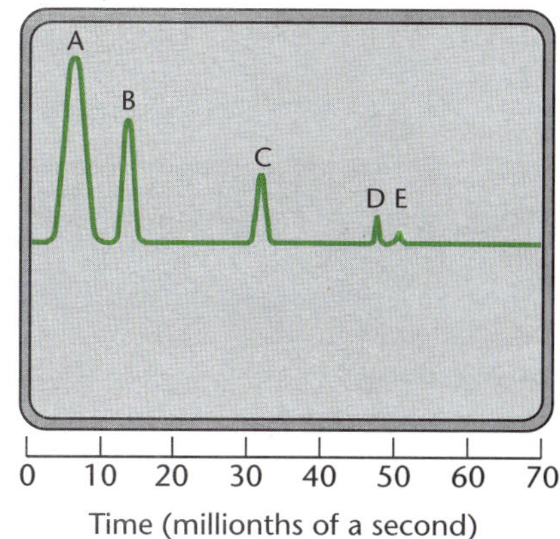

The speed of ultrasound through fat is 1000 m/s.

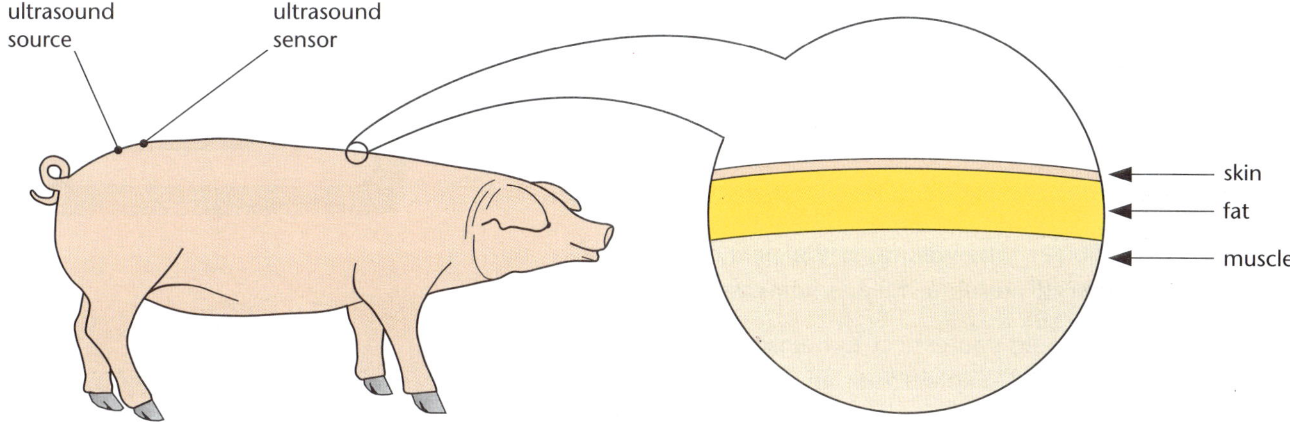

5 (a) Copy and complete each of these equations for radioactive decay.

(b) Some igneous rocks contains a radionuclide with a half-life of 500 000 years.

Rock A contains 1 atom of the radionuclide for every 7 atoms of the decay product.

Rock B contains 1 atom of the radionuclide for every 4 atoms of the decay product.

How old are rocks A and B? [You can calculate A but you will need to draw a graph for B.]

$$^{14}_{6}C \longrightarrow \,^{?}_{?}N + \,^{0}_{-1}e$$

$$\,^{?}_{?}\,\text{particle}$$

$$^{?}_{?}U \longrightarrow \,^{230}_{90}Th + \,^{4}_{2}?$$

$$\alpha\,\text{particle}$$

(c) The diagram shows a different kind of nuclear reaction.

  (i) What is this type of nuclear reaction called?

  (ii) Where is this kind of nuclear reaction used?

  (iii) Write down <u>three</u> ways that the nuclear reaction is different from radioactive disintegration.

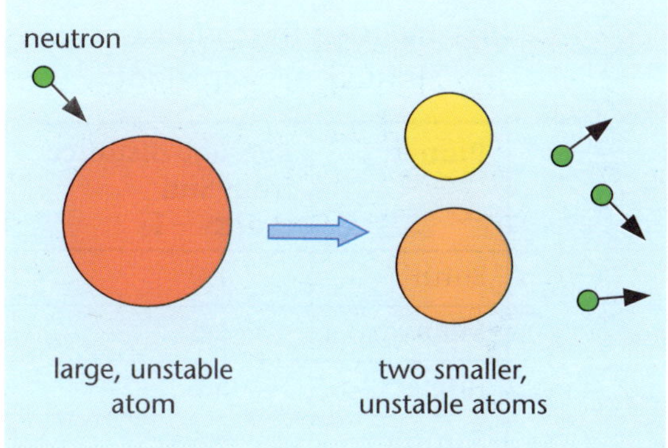

Waves and radiation: Additional questions   **71**

Additional questions: Forces

You may need to use any of the formulas under 'Forces' on pages 55 and 56.

1 The apparatus shown in the diagram is used to measure the volume of a fixed mass of air as the pressure is very slowly increased.

(a) Copy and complete the table.

Pressure (Pa × 100 000)	Volume (cm³)
1	24
1.5	
2	12
10	

(b) The pressure on the air is <u>suddenly</u> reduced to 100 000 Pa. The volume of the air is now 23 cm³. Explain this result as fully as you can.

(c) What would you expect to happen during the next few minutes? Explain your answer.

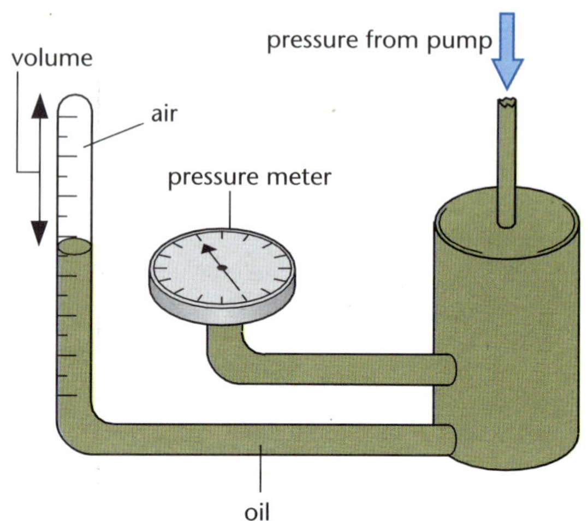

2 Between Mars and Jupiter there is a narrow belt containing thousands of asteroids. The diagram shows some of the largest asteroids.

(a) Write down four ways in which asteroids can be different from each other.

(b) Ceres, the largest asteroid, has an orbital period of 4.6 Earth years. Estimate the average distance of Ceres from the Sun:

  (i) by drawing a graph using the information in the table below;

  (ii) by using the formula
$$[\text{distance}]^3 = [\text{period}]^2$$

Planet	Average distance from Sun (Earth = 1)	Time taken for one orbit (Earth = 1)
Earth	1.0	1.0
Mars	1.5	1.9
Jupiter	5.2	11.9
Saturn	9.6	29.5

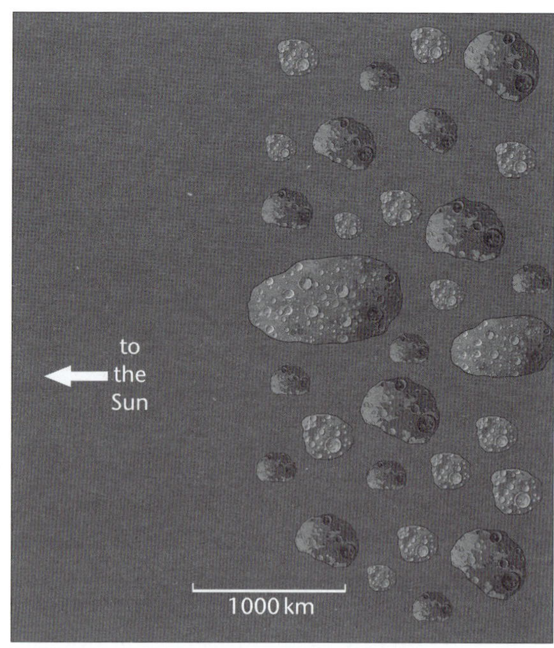

Some of the bigger asteroids. Their different distances from the Sun are shown to the same scale as their sizes. (Note: They are spread around the asteroid belt, not all bunched together as in this diagram.)

3 The diagram shows a hydraulic system.

    (a) Calculate the area of the slave piston.

    (b) Calculate the force on the master piston.

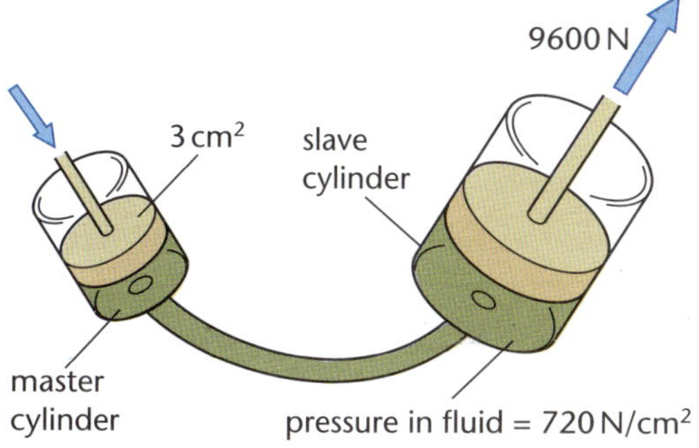

4 Study the graph, then answer the questions below.

    (a) What is happening during parts A, B and C on the graph? Give your reasons.

    (b) What is the speed during part A?

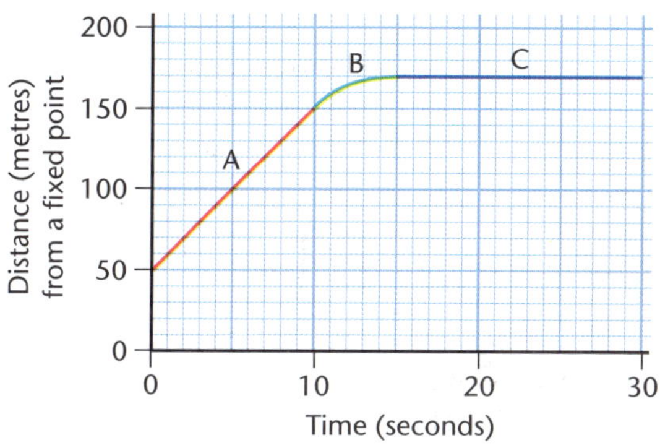

5 Study the graph, then answer the questions below.

    (a) What is the acceleration during parts A, B and C on the graph?

    (b) What is the distance travelled:

        (i) during part B?

        (ii) during part C?

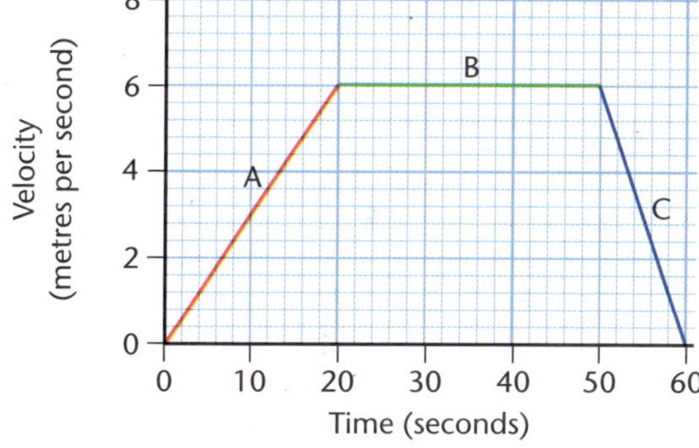

6 Read the extract from a road safety leaflet opposite. Then answer the questions below.

    (a) How many times more kinetic energy does a car have:

        (i) at 30 mph than at 20 mph?

        (ii) at 40 mph than at 20 mph?

    (b) How well do your answers to (a) match the details in the leaflet? Suggest reasons for any discrepancies.

For pedestrians, the speed at which they are hit is critical.

- At 20 m.p.h. only one in twenty is killed. Most injuries are slight and three out of ten suffer no injury at all.

- At 30 m.p.h. nearly half are killed.

- At 40 m.p.h. nearly all are killed.

Car drivers are better protected, but even for them an impact at 40 m.p.h. is five times as likely to result in serious injury as an impact at 20 m.p.h.

Energy

[These Higher-tier questions can be used to replace the 50% of easier questions in *Science Foundations* module tests.]

Q1 Aspects of thermal energy transfer can often be explained in terms of particles.
Use words from the following list to complete spaces 1–4 in the sentences below.

> **conduction**
> **convection**
> **insulation**
> **radiation**

…**1**… is caused by particles <u>not</u> being able to move around.
…**2**… is caused by waves which <u>don't</u> need particles to travel through.
…**3**… is caused by particles of matter moving further apart.
…**4**… is caused by the movement of small charged particles called electrons.

Q2 The table shows some ways of reducing the rate of thermal energy transfer from a room.

	Cost (£)	Saving per year (£)
Draughtproof door	10	25
Thicker carpet	300	75
Double glazing	300	50
Line external wall	200	100
Thicker curtains	75	25

Which <u>two</u> of the methods are the most cost-effective?

Q3 You may find this formula useful when answering part of this question:

energy transferred = power × time
(kilowatt-hours, kW h) (kilowatts, kW) (hours, h)

1 A radio uses 4 cells which cost 50p each. The cells will supply energy at a rate of 0.5 W for 40 hours. How many kilowatt-hours of energy do the cells transfer in the 40 hours?

    A 0.02
    B 0.08
    C 20
    D 80

2 What would 1 kilowatt-hour (1 Unit) of electricity from the cells cost?

    A 10p
    B £6.25
    C £25.00
    D £100.00

3 The cells in the radio last for about a month. Re-chargeable batteries and a solar battery charger cost £36.00. How long would they take to pay for themselves?

    A 6 months
    B 18 months
    C 3 years (36 months)
    D 6 years (72 months)

4 The solar battery charger is quite bulky. It takes about 6 hours in bright sunlight to re-charge the batteries. The charger would be most useful …

    A … on a hiking holiday.
    B … in the UK in winter.
    C … in a remote African village.
    D … in a town where one Unit of mains electricity costs 10p.

Q4 Different energy sources have different advantages and disadvantages.

1 Which of the following is an advantage of renewable energy sources?

  A You can guarantee a supply of energy whenever you need it.
  B Transferring energy to a generator is cheap.
  C They do not produce chemical pollution.
  D They do not affect the environment in any way.

2 Which of the following is a disadvantage of renewable energy sources?

  A They have no fuel costs.
  B They do not produce chemical pollution.
  C The supply of energy does not always match the demand.
  D The energy source is likely to continue for millions of years.

3 Which of the following is a disadvantage of gas-fired power stations?

  A They have a very short start-up time.
  B They add greenhouse gases to the atmosphere.
  C They produce less chemical pollution than other fossil fuels.
  D They can be shut down very quickly.

4 Which of the following is an advantage of nuclear power stations?

  A They produce waste that stays radioactive for thousands of years.
  B They do not produce the gases that cause acid rain.
  C They take a long time to shut down.
  D They have very high de-commissioning costs.

Q5 You may find these formulas useful when answering parts of this question:

$$\text{change in gravitational potential energy (J, joules)} = \text{weight (N, newtons)} \times \text{change in height (m, metres)}$$

$$\text{power (W, watts)} = \frac{\text{energy transferred (J, joules)}}{\text{time taken (s, seconds)}}$$

$$\text{efficiency} = \frac{\text{total energy transferred}}{\text{energy usefully transferred}}$$

An electric motor is used to raise a 0.5 kg mass.

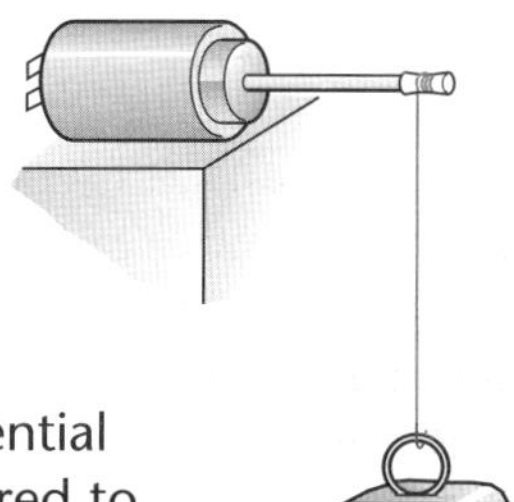

1 How much gravitational potential energy is transferred to the mass when it is lifted 0.5 metres above the ground?

  A 0.25 J   **C** 2.5 J
  B 1 J   **D** 10 J

2 It takes 5 seconds to lift the mass. What is the output power of the motor?

  A 0.2 W   **C** 5 W
  B 0.5 W   **D** 12.5 W

3 To lift the mass 0.5 metres, the motor transfers 4 J of electrical energy. What is the efficiency of the motor?

  A 0.125   **C** 0.4
  B 0.25   **D** 0.625

4 When the mass falls back down from 0.5 metres, the motor acts as a generator and produces 1 J of electricity. How efficient is this generator?

  A 0.1   **C** 0.4
  B 0.25   **D** 0.5

Electricity

[These Higher-tier questions can be used to replace the 50% of easier questions in *Science Foundations* module tests.]

Q1 The table shows some of the symbols used in electric circuit diagrams.
Match the words in the list with the numbers 1-4 in the table.

diode   **thermistor**
LDR   **variable resistor**

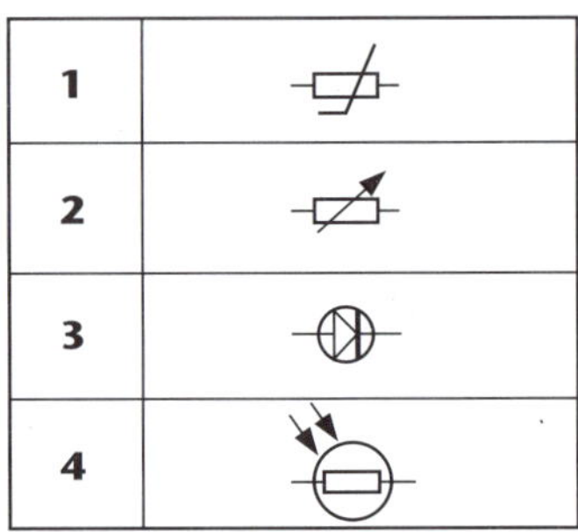

Q2 The diagram shows the electrolysis of some acidified water.

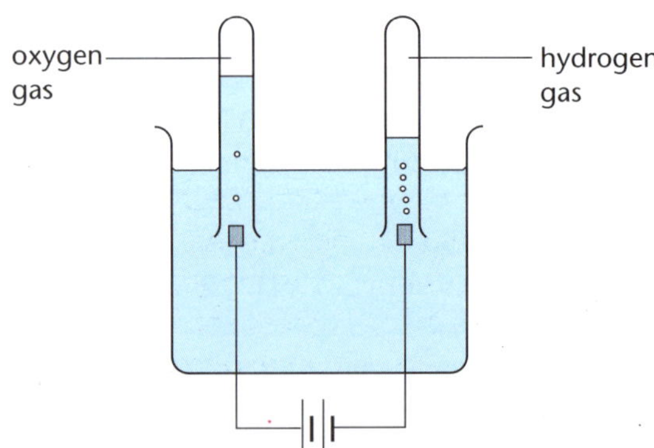

On which <u>two</u> of the following does the volume of the oxygen and the hydrogen that is produced depend?

A only on the amount of charge that flows
B only on the time that a current flows
C only on the size of the current
D on both the size of the current and the potential difference
E on both the size of the current and the time that it flows

Q3 You may find this formula useful when answering part of this question:

$$\underset{\text{(J, joules)}}{\text{energy transferred}} = \underset{\text{(V, volts)}}{\text{potential difference}} \times \underset{\text{(C, coulomb)}}{\text{charge}}$$

1 During an electrolysis, a potential difference of 4.6 V is used to produce a current of 0.92 A. How much charge flows when the current is switched on for 2 minutes?

A 1.84 coulombs
B 9.2 coulombs
C 110.4 coulombs
D 552 coulombs

2 How much energy is transferred each second?

A 4.232 J
B 4.6 J
C 5.0 J
D 9.0 J

3 The 4.6 V d.c. supply used for the electrolysis is produced from the 230 V a.c. mains. The a.c. voltage is first reduced to 4.6 V. The primary of the transformer that is used has 1000 turns. How many turns has the secondary?

A 4.3
B 20
C 50
D 50 000

4 What will the current be in the primary? (You can assume that there are no energy losses in the transformer.)

A 0.004 A
B 0.0184 A
C 0.92 A
D 46 A

Q4 The graph shows the current through component X when different potential differences are applied across it.

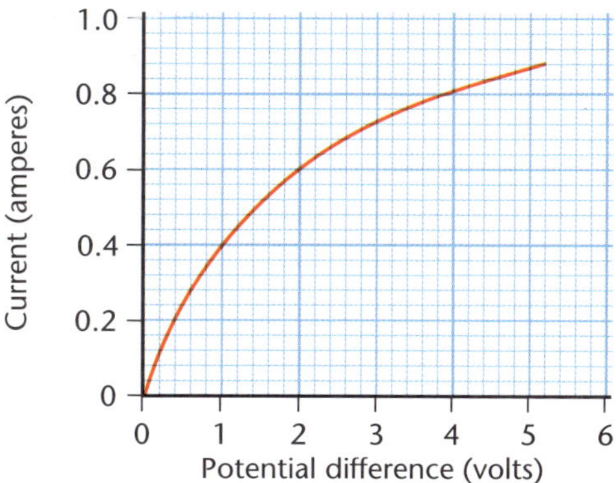

1 Component X is …

   A a diode
   B a filament lamp
   C a light dependent resistor
   D a thermistor

2 When 4.0 V is applied across it, the resistance of component X is …

   A 0.2 ohms
   B 3.2 ohms
   C 4.8 ohms
   D 5.0 ohms

3 When 4.0 V is applied across component X, the rate at which it transfers energy is …

   A 3.2 J
   B 3.2 W
   C 5.0 J
   D 5.0 W

4 Which of the following best describes what happens to component X as the potential difference across it is increased?

	Temperature	Resistance
A	increases	increases
B	increases	decreases
C	decreases	increases
D	decreases	decreases

Q5 The diagram shows a model a.c. generator.

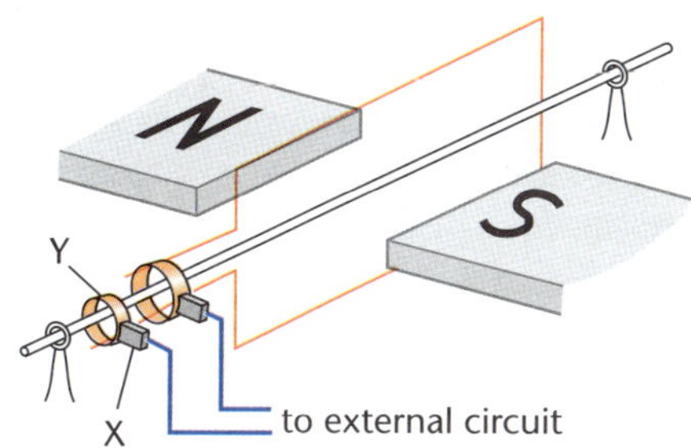

1 Label X shows …

   A an armature
   B a brush
   C a commutator
   D a slip ring

2 Label Y shows …

   A an armature
   B a brush
   C a commutator
   D a slip ring

3 The job of X and Y together is …

   A to keep reversing the direction of the induced potential difference
   B to increase the size of the induced potential difference
   C to connect the induced potential difference across an external circuit
   D to connect an external potential difference across the coil

4 The coil in the generator is changed for one with a bigger area, but still with the same number of turns and made from the same wire. The coil is then rotated between the same magnets at the same speed as before. Which of the following is now true?

	Resistance of coil	Induced p.d. across coil
A	decreased	decreased
B	decreased	increased
C	increased	decreased
D	increased	increased

Glossary/index

This Glossary/index includes terms which are newly introduced in *Science Foundations Extension* or which have different examples of their use than in *Science Foundations*.

acceleration: the rate at which the *speed* or *velocity* of a moving object changes; can be calculated from the *gradient* of a *velocity–time* graph   **44–45**

alpha (α): one of the types of *radiation* emitted when the *nucleus* of a *radionuclide decays*; it consists of particles each of which is the *nucleus* of a helium atom **32–33, 35**

atmosphere: a unit of *pressure*; it is 100 000 *pascals*, the average pressure of the Earth's atmosphere at sea-level **36**

base-load demand: the minimum amount of mains electricity that needs to be generated all the time   **12**

beta (β): one of the types of *radiation* emitted when the nucleus of a *radionuclide decays*; it consists of particles called *electrons*   **32–33, 35**

big bang: the name of the theory which says that the *Universe* started by rapidly expanding from a single point **48–49**

brushes: these rub against the *slip rings* in a *generator* so that the *potential difference* produced can be applied to an external circuit   **16**

C: short for *coulomb*

capital cost: the cost of the materials, equipment and labour when setting up, for example, a power station   **10**

chain reaction: a reaction, such as a *nuclear fission*, that continues by itself once it has been started   **35**

charges: electrical charges can be *positive* or *negative*; static electricity is caused by a surplus of one type of charge; electric *currents* are caused by moving charges; the unit of charge is the *coulomb*   **20–21, 23**

comet: a small body which moves round the *Sun* in a very *elliptical orbit* with a long *period*   **41**

conduction: the transfer of *thermal energy* by a material without the material itself moving   **6**

conductor: (1) a material which readily allows the *conduction* of *thermal energy*   **6**
(2) a material which readily allows an electric *current* to pass through it.

convection (current): the transfer of *thermal energy* by a hot liquid or gas itself moving   **6–7**

cost-effective: something that more than repays the cost of setting it up; things that are cost-effective usually have a short *pay-back time*   **8**

coulomb: the unit of electrical *charge*; the amount of charge that flows when a *current* of 1 ampere flows for 1 second; *C*, for short   **21**

current: the movement of electrical *charges* caused by a *potential difference*   **14–15, 18, 20–23**

decay: what happens to the unstable *nucleus* of a *radionuclide* atom when it emits *radiation*   **33, 34**

de-commissioning: the process of dismantling, for example, a power station when its useful life is over   **11**

diffraction: the bending of waves by the edge of a barrier   **30–31**

distance–time graph: a graph of the distance moved by an object plotted against time; the *gradient* of the graph represents *speed*   **44–45**

earth(ing): connecting something to the Earth with an electrical *conductor*   **19**

efficiency: the fraction of the energy supplied to a device that it transfers in the form that we want   **12–13**

efficient: a device which transfers a large proportion of the energy supplied to it as the form of energy we want (or a bigger proportion than other devices do)   **12–13**

electrolysis: splitting up a molten or dissolved chemical compound by passing an electric *current* through it   **21**

electromagnetic waves: how energy is transferred by the types of *radiation* that belong to the electromagnetic spectrum (radio, microwave, infrared, light, ultraviolet, X-rays, *gamma* rays)   **25**

electron: one of the types of particle from which atoms are built; they carry a tiny *negative charge*   **6, 20, 32–33**

ellipse, elliptical: an oval shape; *orbits* are often elliptical   **41**

escape speed: the *speed* that something needs to have to be able to escape the *gravity* of another object (such as the Earth)   **47**

fission: see *nuclear fission*

force multiplier: a device, such as a *hydraulic system*, for obtaining a large output force from a small applied force   **38**

frequency: the number of complete vibrations or oscillations each second; frequencies are measured in *hertz*   **25**

fuse: a device which melts and so breaks a circuit that has too big a *current* flowing through it   **19**

fusion: see *nuclear fusion*

galaxy: a local group containing billions of *stars*; the *Universe* is made up of billions of galaxies   **48**

gamma (γ): one of the types of *radiation* emitted when the nucleus of a *radionuclide decays*; it consists of very short-wave, high *frequency electromagnetic waves*   **32–33, 35**

generator: a device which produces a *potential difference* from coils of wire moving in a magnetic field (or vice-versa)   **16**

gradient: the slope of a graph; on a *distance–time* graph the gradient represents *speed*; on a *velocity–time graph* the gradient represents *acceleration*   **44–45**

gravitational potential energy: the energy that is transferred to an object when it is moved against the force of *gravity*   **12–13**

gravity: the force between two objects because of their mass; for this force to be big enough to be noticeable, at least one of the bodies must have a very large mass   **40, 42**

half-life: the time that it takes for the *radiation* emitted by a sample of a *radionuclide* to fall to half its original level (or for half of the unstable radionuclide atoms to *decay*)   **33, 34–35**

hertz: the unit of *frequency*; the number of complete vibrations or oscillations each second; *Hz*, for short   **25**

hydraulic system: a system that uses *pressure* in a liquid to transmit forces that are the required size and direction to where they are needed   **38–39**

Hz: short for *hertz*

ion: particles in a molten or dissolved ionic compound that have an electric *charge* and allow the *conduction* of an electric *current*   **20**

kinetic energy: the energy that an object has because it is moving   **7, 46–47**

LDR: see *light dependent resistor*

light dependent resistor: an electrical component whose *resistance* decreases as the intensity of the light that falls on it increases   **15**

light-year: the distance that light travels through space in a year; it is used as a unit for the very large distances to *stars* and *galaxies*   **48**

live: one of the two wires in the mains supply that carries the *current*; it has a varying *potential difference* relative to the *neutral* and *earth* wires; in the flex connected to electrical appliances the live wire is brown   **19**

longitudinal: waves in which the vibrations are <u>along</u> the same direction as the wave is travelling   **24–25**

medium: the general name for a substance through which a wave is travelling; the plural is 'media'.

negative: one of the two types of electrical *charge*; electrons carry a tiny negative charge   **20**

neutral: one of the two wires in the mains supply that carries the *current*; it has hardly any *potential difference* relative to the *earth* wire; in the flex connected to electrical appliances the neutral wire is blue   **19**

neutron: one of the particles found in the *nucleus* of atoms; it has the same mass as a *proton* but has no electrical *charge*   **33, 35**

neutron star: a *star* that remains after a *supernova* explosion; it is made of matter that is even denser than the matter in a *white dwarf*   **43**

nuclear equation: an equation that shows what happens to the *protons* and *neutrons* in the *nucleus* of a *radionuclide* atom when it *decays*   **33**

nuclear fission: the deliberate splitting of an unstable *radionuclide nucleus* by bombarding it with *neutrons*; once it has started, this is a *chain reaction*   **35**

nuclear fusion: the nuclear reaction that occurs in *stars* when small atomic *nuclei* join together (fuse) to form larger ones   **42–43**

nucleus: the central part of an atom; it is made from *protons* and *neutrons*; the plural is 'nuclei'   **33, 35**

ohm: the unit of electrical *resistance*; Ω, for short   **14**

orbit: the circular or *elliptical* path of a smaller object as it moves around a larger one, such as a *planet* around the *Sun* or a *satellite* around a planet   **40–41**

Pa: short for *pascal*

pascal: a unit of *pressure*; 1 pascal is 1 newton per square metre; *Pa*, for short   **36**

pay-back time: the time it takes for something to pay for its cost by the money it saves   **8**

period: the time it takes for any regular motion, such as an *orbit*, to go through one complete cycle   **40**

p.d.: see *potential difference*

planet: a large object, such as the Earth, that is in *orbit* around the *Sun*   **40–41**

positive: one of the two types of electrical *charge*   **20**

potential difference: the difference between two points that causes an electric *current* to flow between them; units of potential difference are *volts*; potential difference is often shortened to p.d. or called voltage   **14–15, 18, 22–23**

potential energy: see *gravitational potential energy*

power: the rate of transfer of energy; units are joules per second (J/s) or watts (W)   **18, 23**